The Journey of Genetic Information: How DNA Guides Cellular Function

Furguson

Contents

1 Introduction 1

   1.1 Gene expression data analysis...1

   1.2 RNA-Seq for gene expression measurement.......................................2

   1.3 RNA-Seq analysis workflow..2

   1.4 Curse of dimensionality 4

   1.5 Unsupervised learning on genomics data 5

   1.6 Motivation...7

   1.7 Aims and Contributions 8

2 Dimensionality Reduction and Clustering 10

   2.1 Linear Feature Extraction Methods..12

      2.1.1 SVD...12

      2.1.2 PCA...13

      2.1.3 Random Projection 14

   2.2 Nonlinear Dimensionality Reduction Methods.......................................15

      2.2.1 Multidimensional Scaling...15

          2.2.1.1 Classical multidimensional scaling.............................16

          2.2.1.2 Non-Metric Multidimensional Scaling........................16

2.2.2 Isomap 17

2.2.3 LLE 18

2.2.4 t-SNE 18

2.2.5 UMAP 19

2.2.6 Hierarchical Clustering 23

2.2.7 K-means Clustering..23

3 Tuning Hyperparameters **25**

3.1 General approach and distance measure applied 25

3.2 Specific adopted approach..26

3.3 Relevant hyperparameter considerations for each method.........................27

4 K-means Clustering-based Evaluation **29**

4.1 Framework...29

4.2 Classification Metrics 32

4.2.1 Accuracy 32

4.2.2 Precision 32

4.2.3. Recall..32

4.2.4 F-score...32

4.2.5 Cohen's Kappa...33

4.3 Clustering validity metrics 33

4.3.1 Adjusted Rand Index..33

4.3.2 Homogeneity 33

4.3.3 Completeness 34

4.3.4 V-measure 34

4.3.5 Silhouette coefficient 34

4.3.6 Fowlkes-Mallow Score...35

5 GTEX Dataset..**37**

5.1 Data source and pre-processing 37

5.2 Results..38

5.2.1 Qualitative evaluation results...38

5.2.2 Quantitative evaluation results 46

5.2.3 Summary 50

6 Single-cell Dataset 51

6.1 Single-cell data particularities 51

6.2 Data source and pre-processing 51

6.3 Results 52

6.3.1 Qualitative evaluation results 53

6.3.2 Quantitative evaluation results 61

6.3.3 Summary 65

7 Computational Demands 66

8 Conclusions 68

8.1 Summary on Guideline Recommendations 68

8.2 Main Contribution 70

8.3 Future Work 71

A Supplementary Visualizations of GTEx dataset 1

B Supplementary visualizations of scRNA-Seq dataset 25

Chapter 1

Introduction

1.1　Gene expression data analysis

As the primary unit of life, cells constitute a complex system that contain the information needed to regulate itself by a set of organic molecules in continuous interaction. It relies on fundamental components for functional and structural tasks, such as ribonucleic acids (RNA) and proteins. The central dogma of molecular biology, first proposed in 1958 by Crick, states that the coding information contained in Deoxyribonucleic acid (DNA) is sequentially transferred to produce proteins, in a flow which involves an intermediate messenger – RNA. The portions of the DNA that comprise the instructions to produce proteins are an object of continuous interest and study, which are commonly known as genes. This process involves DNA replication, transcription into mRNA (messenger RNA), and a posterior translation of the information into a chain of amino acids.

It is known that, normally, the same genetic information is observed in every cell of a multicellular organism. Thus, a question that may arise, is how differentiated cells have such a specific functioning. The distinction between the various cell types is generated by the different quantities of mRNA produced by the various genes, or, in other words, its specific gene expression. The collected expression levels of a sample across the various genes are defined as its expression profile. This profile may then be considered a specific and relevant signature for the different cell and tissue types. Since the first observation of multicellular organisms by Robert Hooke in 1665, cataloguing and classifying the different cell types that multicellular organisms comprise has also been an object of great attention and curiosity, which extends to present times. The avid search for answers to various questions related to the origin, functionality, diversity, quantity, and interactivity in a tissue of the various cell types still fascinates us and, even though much knowledge has already been learned, much is yet unknown.

The flow of genetic information occurs by an interconnected and dependent expression of various genes, by a positive or negative feedback of the various gene products. A better knowledge on this process enables a better understanding of the functioning of healthy tissues and, naturally, on the origin and functioning of various diseases [1][2][3][4][5][6]. However, there is a lack of sufficiently comprehensive studies with various cell types and individuals that may enable a broader and deeper understanding on this extremely relevant topic.

1.2   RNA-Seq for gene expression measurement

A large fraction of what we know about RNA is based on biochemical techniques, in which a small quantity of specific molecules is examined. In fact, high-throughput methods only emerged in the mid-1990s.

With the introduction of next-generation sequencing on the first decade of this millennium [7], RNA sequencing (RNA-Seq) emerged as a new methodology that uses these technologies to overcome previous challenges on RNA analysis, such as low specificity or sensitivity for some genes. Contrarily to anterior techniques, this single-pair resolution method directly determines the complementary DNA (cDNA) sequence. In addition to providing a considerably more accurate measurement of transcript quantities and structure in a short amount of time, it remarkably permitted the study of unknown genes and novel transcripts isoforms, by unravelling sequence identity. As a result, it allowed a much better knowledge on genomic functionality which has, naturally, revolutionized biology and medicine research. Since its development, it has been the standard method of election for gene expression analysis. An unprecedented understanding of the transcriptome for diverse cell types and species has been unravelled since then. However, adequate methods for this information analysis still need development and improvement [8].

1.3   RNA-Seq analysis workflow

Subsequently to the selection of high-quality reads from high throughput sequencing by a quality control step, the data analysis follows. The initial stage consists in aligning the resulting reads to a reference genome or transcript annotation if existent. Otherwise, the reads are aligned with a *de novo* assembly, generating a set of sequences that contain the captured transcriptional repertoire. If a genome and gene annotation is available for the organism in study, several tools must be applied to count the number of mapped reads per gene. This is a two-step procedure that includes mapping the reads against the reference sequences.

Programs like TopHat [9], Bowtie [10][11], Star [12] can be used. The second step consists of counting the number of reads per feature, *i.e.*, genes, transcripts, or exons. HTSeq [13] and featureCounts [14] are candidate programs for this task.

After this alignment stage, the level of expression for each gene is quantified. As the number of mapped read counts depends on several factors, including sequencing depth and gene length, it is of crucial importance to proceed with a normalization step with the aim of reducing RNA-Seq non-biologically derived inherent variation and allow intra and inter-sample gene comparison. Two of the most used measures to perform this normalization step are RPKM, standing for reads per kilobase per million mapped reads and TPM, standing for transcripts per kilobase per million mapped reads. These measures are used for the normalization of the datasets analysed in this thesis, and both seek to normalize the data for sequencing depth and gene length. The main distinction between these metrics' procedure is that, in TPM, the gene length is first normalized prior to the sequence depth normalization, whereas in RPKM the reverse is performed. This distinction makes the sum of all TPMs in each sample or cell to be equal, which facilitates the comparison of the proportion of reads.

After gene expression quantification, a high-level analysis follows, with the objective of extracting relevant information and biological patterns. This analysis consists in mapping the data by a particular representation process, which should be selected according to the type of data structure to be analysed. Machine learning methods such as clustering, DR and classification are often applied, as well as non-learning methods such as differential expression analysis or gene network inference. In this thesis, we focus on the DR method. This method aims to overcome the problems related to a high number of features to a low number of samples, which is now discussed in the following section.

1.4   Curse of dimensionality

Due to the advancement in sequencing technology, an exponentially larger volume of data has been produced in Genomics, including diverse genomes. However, genomics data is usually highly dimensional due to the unbalanced combination of a very-high number of features, typically in the thousands, for a relatively smaller number of samples. This imbalance results in increased challenges and obstacles for the proper examination of data, which often may lead to a less reliable identification of patterns and statistical analysis by inference methods, and a lower performance of machine learning algorithms [15][16][17][18]. This imbalance, commonly known as curse of dimensionality, was first introduced by Bellman. It implies that the number of samples needed for a reliable statistical analysis grows exponentially with the number of input variables [19]. Adding extra dimension to a Euclidean space is associated with an exponential increase in each data vector space volume, leading to high data sparsity. In Figure 2, this challenge is illustrated. It is known that a volume of about 78% of the square is covered by the circle (left). However, the largest sphere in a cube covers only about 52%. As can be observed at the right of this figure, this volume, represented by the fraction of points in the hypersphere, may reduce exponentially with the increment of the number of dimensions, with a value close to 0 for 10 dimensions. Hence, the number of samples required to explain statistical analysis would have to grow exponentially with the number of variables. This is normally unfeasible, so, in order to gain valuable insights of complex datasets such as genomics datasets, adequate computational techniques are required, such as DR methods, which may help to capture and visualize significant relations.

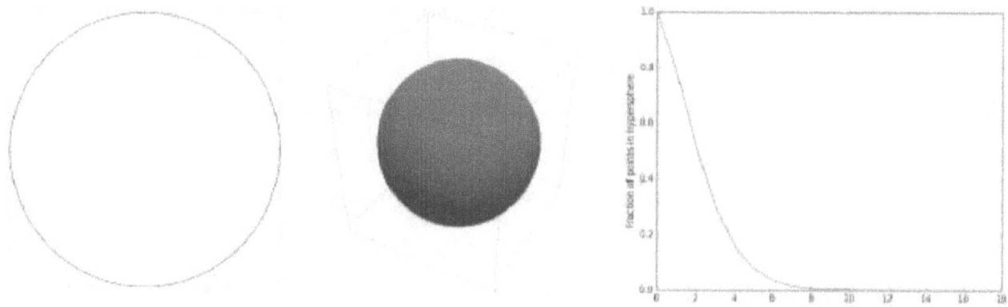

Figure 2 - The curse of dimensionality. Increasing dimensions to a Euclidean space is associated with an exponential increase in volume.

Additionally, the application of these methods diminishes the prior large amount of computations and memory needed for further analysis by projecting the original dimensions into a new set of non-redundant ones, which should consist in a much lower-dimensional subspace, retaining the most important and useful data. There have been abundant examples demonstrating that low-dimension projections can be a very useful help in understanding and visualizing the underlying biology, by summarizing the main characteristics of data [20][21][22][23]. These projections are essential to overcome the aforementioned challenge. Several methods that use multivariate statistical analysis to reduce data dimensionality have been evolving in machine learning methods. These methods have been applied in multiple fields since their first rise in the 20th century, as they have proven to be of indispensable value in order to handle large datasets [15][24]. They can be split into two categories – supervised and unsupervised learning.

1.5   Unsupervised learning on genomics data

Supervised learning aims to identify a mathematical relation between the response variable and the labelled data, which are the predictors. In classification, the learning of this relation permits the prediction of the categories that new input data belongs to, and it is performed by the introduction of representative training examples. This latter technique is used in a posterior stage of study to explain and validate the discovered patterns. The class label might, for instance, consist in a gene biological function, or in a disease subtype, which may be especially useful for research on personalized medicine [25][26].

Unsupervised learning is a widely used machine learning technique for the exploratory data analysis stage which does not directly use class labels. This information may be posteriorly used to verify and compare results, but not in the learning stage. Its main aim is to automatically discover hidden and relevant relations and patterns. This is of crucial importance to adequately visualize data general structure and diminish the computational burdens for the subsequent analysis, as a direct analysis on the reduced attributes space instead of the full initial one is allowed. In this extent, the output of unsupervised techniques can serve as input to supervised techniques.

DR has been a widely used unsupervised learning method for the exploratory data analysis of genomics data. In 2008, Lee et al stated an improved efficacy by the non-linear DR techniques in visualization and classification of gene and protein expression datasets. For this evaluation, the discriminability of supervised classifiers and cluster validity measures in the low-dimensional space were used [27]. Accordingly, in a 2010 comparative study of DR techniques for the visualization of microarray gene expression data, non-linear methods also demonstrated superior performance in a systematic benchmark. This evaluation comprised a combination of Support Vector Machine classification, cluster validation and noise evaluations [28].

After the emergence of single-cell RNA sequencing in 2009, large volumes of this type of data have been generated. Unsupervised clustering has been a crucial step to define possible cell types based on transcriptome similarity. This step allowed the creation of Human Cell Atlas, along with various other atlas projects. Their references can be very useful for pathologic studies, as well as in understanding the basic biology.

Yet, the used methods should have good reliability, and criteria on what constitutes a cell type must be well-established in order for an atlas to have useful application [29]. It is necessary to evaluate which specific methods are more adequate for each data type and aim of analysis. In a 2016 study [30], it was verified that DR methods significantly improved later analysis of data on patient classification and gene clustering. They pointed out, however, that the choice of method to use should be made accordingly to the specific purpose of the analysis, and that for this condition more studies were needed to address it.

Recently in 2019, Ho-Sik Seok et al. [31] compared the effects of various DR methods on gene expression data from the database of Genotype Tissue-Expression (GTEx). The adopted algorithms included local linear embedding (LLE), non-metric multidimensional scaling (NMDS), Principal Component Analysis (PCA), spectral embedding (SE), and t-distributed stochastic neighbour embedding (t-SNE). They reported that each of these methods produced very distinct projections, and that the subsequent DR clustering-based evaluation results were also very different, finally emphasizing the need to choose a method which is more adequate to the specific type of genetic dataset.

Thus, machine learning, and particularly unsupervised learning algorithms, have been indispensable in genomics, by finding patterns that would not be detected otherwise, and enabling a more efficient data analysis. This pattern detection may allow a better understanding of the basic biological mechanisms and facilitate future research of bioinformatics [25][32][33]. Given the absence of corresponding labels, interpretation of the results must be carried in light of the biological context of the data.

1.6 Motivation

A well-appreciated concept in computer science, called "no free lunch theorem", states that all methodological choices have trade-offs. Finding a method that is "best for all" may be impossible, as no algorithm is able to achieve the whole range of desired properties. Therefore, it is very important to evaluate them in a wide set of criteria [34][35]. This observation resulted in the proposal of a multitude of unsupervised learning algorithms over the past half-century. Importantly, various studies have already demonstrated that non-linear methods present better adequacy for complex artificial datasets. Yet, the outcomes on natural datasets are less compelling [36]. Recently in 2019, Kiselev et al. presented a review about the challenges in unsupervised clustering of single-cell RNA-Seq data and stated that, in addition to developing methods for an adequate clustering, there is also a need for methods that may facilitate biological interpretation and annotation [29]. As biological samples have several levels of functional specialization, many authors have implemented the tactic of subsequently re-cluster large clusters identified in an initial stage of clustering in order to deal with an inferior performance when trying to better distinguish rare cell types [37][38]. However, the decision of when it is appropriate to proceed in re-clustering is often complex to take. While clustering large datasets can take a few hours, visualizing and interpreting the outcomes may be more intricate. Furthermore, due to the current vagueness in guidelines for choosing the hyperparameters, there is a great possibility of having a wide range of results. Therefore, it is very important to properly evaluate and compare different DR techniques in genomics datasets, identifying their main advantages and disadvantages. This permits to establish the most adequate procedures to obtain more robust posterior analysis and conclusions.

In the present work, the evaluation of these methods was made in terms of biological interpretability and computational demands, including CPU time and memory requirements. The biological interpretability evaluation and comparison were guided by the biological expectations, such as correct clustering by tissue or cell type (the latter in case of single-cell data). For a more precise and specific quantification, k-means clustering performance was measured subsequently to the implementation of each method. The metrics applied for this measurement were the Adjusted Rand index, Homogeneity, Completeness, V-measure, Fowlkes-Mallows scores, and Silhouette Coefficient.

Hence, this thesis aims precisely to address the emphasized remarks by assessing a larger set of DR methods. To achieve this, a clustering-based evaluation approach combined with a broad range of evaluation measures was applied.

1.7 Aims and Contributions

For a consistent, reproducible, and reliable work, we developed a pipeline and established a framework with precise procedures to properly evaluate and compare the various DR techniques. We started by reviewing the literature for unsupervised learning methods and selecting challenging and adequate gene expression datasets. In a next step, the application of each method, visualizations and qualitative and quantitative comparisons of the results were performed. Various linear techniques were applied, including PCA, Singular Value Decomposition (SVD) and Random Projection (RP), and non-linear techniques such as Isomap (IM), Multidimensional Scaling (MDS) including CMDS (Classical MDS) and NMDS, LLE, t-SNE and the recently developed Uniform Manifold Approximation and Projection (UMAP). Additionally, Hierarchical Clustering (HC) was also implemented. These methods are amongst the most applied for this type of data.

Finally, the analysis of the comparisons for each dataset were produced with recommendations to apply the studied methods and compare them in the most appropriate way. An automatic pipeline development was of crucial importance for an efficient analysis, as it was necessary to repeat it multiple times (Figure 3). The entire work was developed in R programming language in a server with *AMD EPYC 7501* 32-Core Processor and 128Gb of RAM memory.

DR is frequently performed to visualize high-dimensional data in two or three dimensions. Therefore, we focused on the methods' performance on the first four dimensions. Additionally, fixing the number of dimensions allows for a fair comparison of the various feature extraction methods. The inherent capability of noise reduction of the different feature extraction methods was also assessed by comparing different scores to the whole data without any projection.

Thus, to develop the extended framework and pipeline, we considered various relevant standpoints and determined the main aims for this work. These were, namely:

(a) Generate visualizations and compare solutions subsequently to each method application. Appropriate visualization should be generated to properly identify the produced patterns and to compare and interpret in biological terms.

(b) Quantitative evaluation and comparison of the various methods by the observation of k-means clustering performance on their reduced spaces. This performance was measured by the most used classification metrics, and by six clustering validity metrics. For the assessment with classification metrics, a specific contingency table methodology was developed, due to the unsupervised nature of clustering.

(c)   Generate recommendations on the most appropriate procedures to adopt. With all the visualizations and quantitative results generated, guidelines were formulated.

In summary, this project advanced the state of the art by i) applying a larger set of DR methods than previous similar studies; ii) implementing a novel evaluation system procedure, including a more comprehensive number of evaluation metrics. Additionally, a specific hyperparameter-tuning approach was proposed, and a greater number of dimensions were analysed. These expansions provided a further detailed evaluation and comparison framework for DR methods.

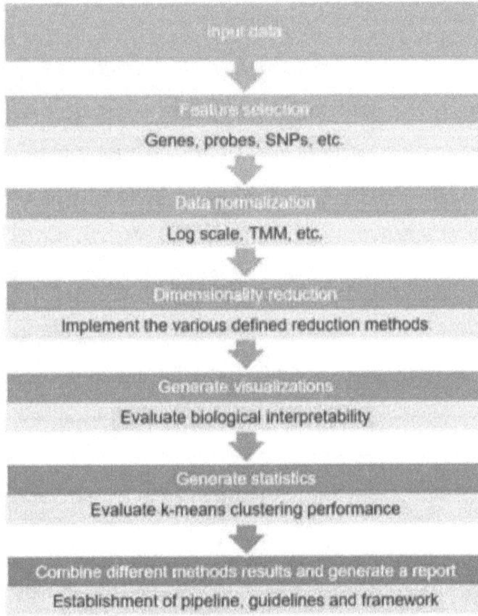

Figure 3 - Project data flow pipeline. With the selection of input data (RNA-Seq dataset), pre-processing steps followed. After this procedure, various DR methods were applied, which resulted in visualizations and statistics for each one. These results allowed the establishment of guideline structures.

Chapter 2

Dimensionality Reduction and Clustering

In gene expression data analysis, DR is applied with the intention of projecting each sample or cell high-dimensional expression profile into a low-dimensional representation. With this method, a transformation or elimination of less informative data is performed. This enables an adequate hypothesis generation and testing. Two main methodologies may be distinguished in DR - feature selection and feature extraction. While feature selection completely discards less informative features, feature extraction performs a feature transformation by constructing new informative and non-redundant features as a combination of the initial ones. In this thesis, we focused on feature extraction algorithms. Nevertheless, a simple feature-filtering procedure is also applied as a pre-processing step on the GTEx dataset (see section 5.1). The selected subset of features followed the Occam's Razor principle, by only keeping the features that are more informative [39][40]. As will be demonstrated, some feature extraction three-dimensional reduced representations still show better results in a clustering-based evaluation than when the full dataset, *i.e.*, without data reduction, is used.

The feature extraction procedure is perfectly conceivable and adequate for genomics data, as distinct genes are frequently correlated if they are involved in the same biological process. Therefore, we can compress multiple genes into a single dimension rather than storing information for each individual gene. This compression is also known as "eigengene" [41]. Considering an input space $\mathbb{R}^D$ with n samples and a target lower-dimensional space $\mathbb{R}^d$ in which $d \ll D$, let $M \in \mathbb{R}^{n \times D}$ be an input dataset of n samples and D features (genes); and $X \in \mathbb{R}^{n \times d}$ its lower-dimensional representation. DR performs a mapping $\Phi: \mathbb{R}^D \to \mathbb{R}^d$ by the optimization of a cost function $\in: \mathbb{R}^d \to \mathbb{R}$ on the reduced space. As a result, important benefits are attained:

- A more efficient subsequent analysis, by creating a new representation of the original features into a lower dimensional subspace, typically smaller than the initial set.

- Noise reduction, as this process averages over several genes.

- Adequate and convenient visualization in a lower dimensional space.

Several algorithms have been proposed, each with its advantages and disadvantages. These may be broadly divided into two main categories: linear and nonlinear. The divergence between these two is shown in Figure 4.

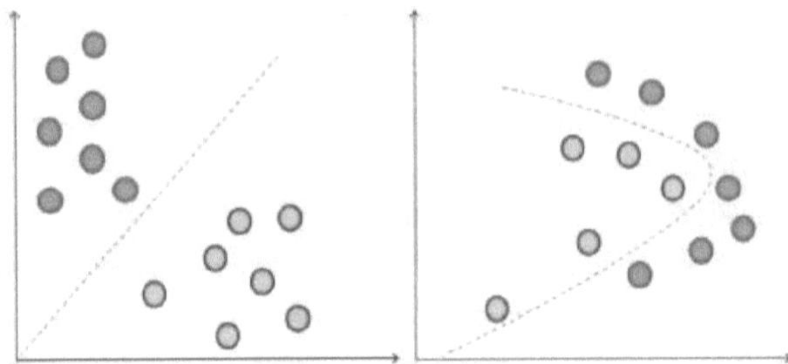

Figure 4 - Representation example of linear (left) and non-linear data structure (right). Non-linear DR methods are much more adequate to capture non-linear data structures.

Clustering is also especially valuable for gene expression data analysis, including the DR process. This method can be applied to discover sets of co-regulated genes or sets of samples/cells with identical genetic expression profiles. It is also applied, for instance, to uncover categories or subcategories of diseased conditions.

Notably, there is a lack of comprehensive studies that evaluate the application of clustering approaches in combination with feature extraction methods, in order to quantitatively assess the efficacy of these important techniques. In this thesis, clustering was applied in two different approaches: a global examination of data structure using hierarchical clustering; and an evaluation approach of the various DR projections, using k-means. The latter method is used in this approach as it has already been demonstrated that, for this type of evaluation, it is generally more effective than the hierarchical clustering method [42][43]. In the following sections, k-means and hierarchical clustering algorithms are overviewed.

In the following sections, an overview on the main functional and mathematical properties of the various feature extraction and clustering methods applied in this thesis is presented.

2.1   Linear Feature Extraction Methods

In a linear DR method, the mapping can be described by a linear transformation. Generally, linear techniques are faster, more robust, and more interpretable than non-linear ones. They have been the most preferred for examining high dimensional data as they typically have a simple implementation procedure and geometric interpretation. Importantly, in contrast to non-linear methods, their low dimensional axes are meaningful, as they are linear combinations of the original axes. In consequence, the resulting features are directly relatable to the input features. These combinations may also give rise to valuable interpretations related to the specific domain in analysis. Furthermore, their computational complexity is usually low both in time and space, an important trait for the analysis of large data, such as the genomics data.

2.1.1   SVD

Singular value decomposition (SVD) [44] is a factorization method of a real matrix into a product of three matrices, U, S and V. The orthogonal and unitary matrices U and V comprise the left and right singular vectors of the resulting matrix X, respectively, and the S matrix only contains values in the diagonal which are the singular values of M. These singular values are the result of a square root of the non-zero eigenvalues and are generally ordered in descending order of magnitude. Multiplying a subset of each of these matrices containing the largest singular values returns a reduced dimensionality matrix. In this manner, the method transforms correlated variables into a new set of uncorrelated ones and may identify and order the dimensions for which data points demonstrate the highest variation. For a data matrix with D features and n instances, singular value decomposition is given by:

$$X = USV^{T}$$

In which U, S and V have dimension $D \times D$, $D \times n$, and $n \times n$, respectively.

2.1.2 PCA

PCA [45] is the most widely used feature extraction method, particularly for biological studies and genomics data. It linearly combines correlated variables into a set of new uncorrelated ones named principal components (PCs) - Figure 5. These components represent eigenvalues which are generated by an orthogonal transformation of the initial variables and a ranking with decreased order of variance is produced. The division of each eigenvalue by the sum of all returns the percentage of variance explained by that PC. This way, the first PC corresponds to the highest eigenvalue explaining the highest variance of the total data variance. Given a data matrix M_{nxD}, its covariance matrix C_M is given by:

$$C_M = \frac{1}{n} MM^T$$

In which the diagonal values of the matrix C_M represent the features' variance and the off-diagonal values represent the covariance. The direction that maximizes variance is an eigenvector of the covariance matrix, or of a correlation matrix, if data has been previously standardized. This transformation can be represented as follows:

$$PM = Y$$

where Y is the transformed dataset ($n \ x \ d$), and P is the linear transformation ($d \ x \ d$) represented by a matrix with the eigenvectors as rows. This transformation is performed such that the resulting covariance matrix diagonal values are rank-ordered, and the off-diagonal values are as close to 0 as possible, such that the covariance between variables is as little as possible. This algorithm assumes that data is linear; is distributed in a gaussian distribution; and that high variance corresponds to the most interesting dynamics.

The overall aim of this algorithm is then to reduce dimensionality by discarding some principal components without much information loss and retaining a combination of variables that best reproduce the dataset structure and information.

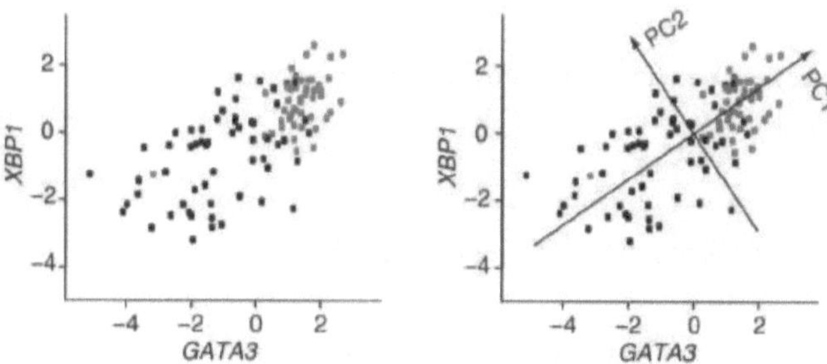

Figure 5 - PCA on a gene expression dataset. (a) Breast cancer samples are plotted according to expression levels for two genes. (b) The two directions along which data contains the largest variance are identified (PC1 and PC2). Taken from [46].

2.1.3 Random Projection

Random Projection reduces data dimensionality in a simple way by projecting the data to a random subspace using a random matrix (R) with unit length columns [47]. A random matrix contains independent and identically distributed random variables. For a Gaussian random projection, each value is sampled in an independent way from a standard Gaussian distribution:

$$R_{ij} \sim N(0,1)$$

The resulting reduced matrix is obtained by my multiplying the original data matrix M by the random matrix R with n rows and d columns:

$$Y = \frac{1}{\sqrt{d}} MR$$

In which the scalar $\frac{1}{\sqrt{d}}$ helps to control with elevated probability that the pairwise Euclidean distance between any two points in the reduced space is very similar to their distance in the original high-dimensional space. Typically, this method performs faster in a high-dimensional dataset than other classical methods such as PCA, as it does not apply eigen-decomposition of matrices. Its key concept is based in the *Johnson-Lindenstrauss Lemma* [48] which declares that distances between data points are approximately preserved if they are projected onto a reduced subspace with sufficient dimensionality. Although various experimental results already demonstrated its capability of successfully preserving distances and structure, there are very few empirical analyses using this technique in gene expression data.

2.2   Nonlinear Dimensionality Reduction Methods

Linear methods, although more capable at preserving global structure [49], do not properly take into account non-linear relationships between variables, which is often an intrinsic characteristic of real data. In genomics, due to the existence of non-linear relations between genes on complex regulatory networks [30], this is often the case.

Nonlinear data structure may be conceptualized as a manifold, in which the distance between two data points over its surface may be larger than their Euclidean distance would imply. These former distances are known as geodesic distances. In other words, points that are distant on a two-dimensional manifold as determined by their geodesic path can seem misleadingly close when measured by their straight-line Euclidean distance, as used in classical linear methods. In this way, they do not reveal the true low-dimensional geometry. Nonlinear DR methods map data along this non-linear manifold by presuming only neighbouring points to be identical enough to be mapped linearly with low error. This high-dimensional space can then be reassembled and mapped back on the lower dimensional space, providing the foundation for a non-linear mapping based on the true distances between any pair of data points.

As a result, these algorithms tend to represent the original data in a more precise way. Generally, and a priori, non-linear transformations present more flexibility, as their outcomes may be easier to interpret by visual inspection and are more attractive in an aesthetical way. The most important limitation of the non-linear methods is the requirement of an optimization step of several input hyperparameters. This can often be a challenging task, considering their high-computational demands. It is generally known that non-optimal hyperparameter configuration may strongly affect results and visualization.

2.2.1   Multidimensional Scaling

Multidimensional scaling (MDS) is a set of methods that perform a simple mapping of the data to a lower subspace and retain the intrinsic properties of the data. Because of this latter aspect, it may be considered a nonlinear DR technique. However, unlike other nonlinear DR methods that we will discuss, an optimization step of the cost function for locality preservation is not implemented for this method.

2.2.1.1　Classical multidimensional scaling

Classical Torgerson's multidimensional scaling (CMDS), also known as Principal coordinate analysis or PCoA, is a metric method that projects data with the consideration of preserving a measure of similarity, which is defined by the distance information between data points [50][51]. This information is computed in a symmetric matrix of dissimilarities based on Euclidean distance. A set of Euclidean distances on n points can be represented exactly in at most n−1 dimensions. The transformation is based on a minimization of an objective function which measures the lack of fit between dissimilarities and the computed distances.

Optimal positions for data points x_i and x_j are obtained by minimization of the least squares error of their calculated pairwise distances, and it is implemented with gradient descent. The minimum value of this objective function is called the stress criterion. The error of a typical stress function of MDS is considered to be the following [52][53]:

$$stress = \sqrt{\frac{\sum(d_{ij} - \hat{d}_{ij})^2}{\sum d_{ij}^2}}$$

In which $\hat{d}_{ij}$ is the predicted distance based on the multidimensional scaling model. This method has been extensively implemented in gene expression datasets, as it is regarded as an effective technique in combination with vector quantization or k-means clustering [54].

2.2.1.2　Non-Metric Multidimensional Scaling

Non-metric MDS conjectures a less strict proximity between the input and the low-dimensional distances, which is, frequently, more realistic to assume. Its basic steps are the following [55], [56]:

1. Initialize the configuration by finding a random low dimensional one or with the classical MDS configuration. The latter was preferred for this project instead of the former.

2. Compute the distances between the data points in the low dimension.

3. Approximate the ranking order between low-dimensional and input distances according to their magnitude. This is performed by seeking optimal monotonic transformation of the input distance, which is attained by isotonic regression, in which a monotonically free-form function is fit.

4. Stress function minimization by reconfiguration of low-dimensional space, moving the points in a way that reduces the discrepancy between the two rankings.

5. Repeat the previous steps from step 2 until convergence.

2.2.2 Isomap

Isomap, or Isometric Feature Mapping, was one of the first algorithms to use a manifold with the aim of unravelling the nonlinear manifold that complex natural observations present. It can be considered an extension of the CMDS algorithm, as it incorporates this linear technique to project high-dimensional data. However, it uses a distinct distance measure methodology. As discussed in the last section, CMDS only considers Euclidean distances between data points, which may result in an inadequate representation of global geometry. Most nonlinear dimensionality approaches focus on local distances and, therefore, use any distance metric. In contrast, Isomap relies on geodesic distances. It was also particularly designed to obtain a global optimal solution and asymptotic convergence, contrarily to other nonlinear DR methods [57]. Its procedure relies on the assumption that the linear combination remains unchanged after linear transformations such as translation, rotation, and scaling. Its implementation may be divided into three steps:

1. Neighbourhood graph development. Firstly, a determination of each data point nearest neighbours by geodesic distances is estimated. The dataset is then represented in a neighbourhood graph, in which shortest distance paths are connected by edges which indicate the weight between neighbouring points.

2. Shortest path computation. The previous constructed graph is transformed with the computed shortest path between pairs of data points. This computation can be performed

by *Floyd-Warshall* algorithm or *Dijkstra*, and a matrix of the conclusive geodesic distances is then constructed.

3. Low-dimensional mapping. Following the global neighbourhood construction, Classical MDS is applied with its cost function to determine the low-dimensional positions of the data points, thereby reducing data dimensionality. With this procedure, this method combines the main algorithmic characteristics of classical techniques, such as MDS and PCA, with the ability to capture nonlinear data structures.

2.2.3 LLE

Locally linear embedding (LLE) is a method that seeks for a set of nearest neighbours for each data vector distributed on a manifold, generating a lower dimensional embedding of a dataset by unfolding it. To perform this search, it uses either the k-nearest neighbours according to the Euclidean distance, or the data points that lie within a fixed distance.

For this reason, it resembles the Isomap algorithm, as it strives to preserve the original data neighbourhood by representing each vector as a linear combination of its neighbours. Additionally, its procedure also relies on the assumption that this linear combination remains unchanged after linear transformations. Yet, in contrast, LLE utilizes two constrained least squares optimisation problems which are quicker optimization methods than Isomap uses for sparse matrices. This combination assigns weight to each neighbour based on the respective reconstruction error, minimizing a cost function. The projection is then finally produced by retrieving the weights that better approximate the initial high-dimensional data [58][59].

2.2.4 t-SNE

t-SNE is a recent algorithm that aims to construct a lower-dimensional configuration based on the other local points to the one being considered, instead of using distance. It was designed to minimize the divergence (*Kullback-Leibler* divergence) between a probability distribution in the high-dimensional space and the one recreated in the lower dimensional space. To measure the pairwise similarity between data points, it uses *Gaussian* distribution for the probability distributions in the original data. The normal distribution reflects the mean as the point location. The conditional probabilities for the high-dimensional space are, thus, calculated as:

$$p_{i|j} = \frac{e^{\frac{-d(x_i,x_j)^2}{2\sigma_i^2}}}{\sum_{k \neq i} e^{\frac{-d(x_i,x_j)^2}{2\sigma_i^2}}}$$

Where x_i, x_j and x_k are points in the high-dimensional space and σ^2 is the variance of normal distribution for a given x_i. The σ^2 is used so that each point has a fixed number of neighbours. The final pairwise probabilities are computed as $p_{ij} = \frac{p_{i|j} + p_{j|i}}{2n}$, where n is the number of samples.

For the recreation of the low-dimensional space probability distribution, the *Student t-distribution* is applied instead, with one degree of freedom:

$$q_{ij} = \frac{(1 + d(i,j)^2)^{-1}}{\sum_{k \neq l} (1 + d(k,l)^2)^{-1}}$$

This specific distribution prevents the points to get overcrowded at the centre, avoiding the effect of the *curse of dimensionality*. The ultimate objective is to make p_{ij} as similar to q_{ij} as possible, in order to approximate the lower-dimensional space to the original one. This cost-function can be computed as:

$$KL(P||Q) = \sum_{j \neq i} p_{ij} \, log \frac{p_{ij}}{q_{ij}}$$

The principle of focusing more on a local than global structure level, hence, permits data points to group more closely to each other, which may improve the visualization outcome. Additionally, it may reveal structures at various scales and data that lies in different manifolds or clusters. The number of nearest neighbours taken into consideration for the projection in the lower dimensional space can be controlled by the *perplexity* hyperparameter. As it is a stochastic method, various runs with unique seeds may result in different projections. Nonetheless, t-SNE becomes a deterministic method if initialized with PCA, a procedure which we adopted for this project, as it significantly enhances this methods' performance.

2.2.5 UMAP

Uniform Manifold Approximation and Projection (UMAP) is a more recent DR algorithm [60], [61] that aims to preserve global structure in addition to preserving as much local structure as t-SNE, and in less computational time. It has already been demonstrated to be remarkably useful for accurately defining cell types of scRNA-Seq experiments [62][63][64][63]. However, there are very few studies of this method on bulk RNA-Seq data yet.

The algorithm uses simple combinatorial building blocks, known as simplices, to compute a low-dimensional object. The reduced topological structure is constructed using low-dimensional simplicial complexes. A low-dimensional simplex is produced in a high dimensional distribution by connecting k-nearest neighbours in a circle, with every point as the centre. Exponential probability distribution representation in used in the following way:

$$p_{ij} = e^{\frac{d(x_i, x_j) - \rho_i}{\sigma_i}}$$

Similarly, to Isomap and t-SNE methods, UMAP uses a k-neighbour based graph, and can be described in two phases. In the first phase, a weighted k-neighbour graph is created, and in the second phase, a reduced layout of this graph is computed:

1. Weighted graph construction

In contrast to t-SNE, UMAP does not necessarily use Euclidean distances to compute exponential probability distribution, but rather any distance metric. Additionally, UMAP does not normalize either high or low-dimensional probabilities. This practice allows for a dramatic reduction in the computing time of the high-dimensional graph, as summation or integration processes are very demanding. The distances between high-dimensional neighbouring points are converted into weights, resulting in a weighted-graph of the high-dimensional data. The yet non-symmetrized weights of the graph in the first phase are obtained by:

$$v_{j|i} = e^{\left[\left(-d(x_i, x_j) - \frac{\rho_i}{\sigma_i}\right)\right]}$$

In which $d(x_i, x_j)$ represents the pairwise distance between high dimensional points i and j, ρ_i is the distance of point i to its nearest neighbour, and σ_i is a normalization term corresponding to the diameter of the nearest neighbour data point of x_i. The computation of ρ ensures the local connectivity of the manifold, and the distance metric varies from point to point, resulting in a locally adaptive approach. Yet, as this is not a symmetric function, the weights need to be symmetrized. This process is also slightly different from the one applied in t-SNE method:

$$v_{ij} = \left(v_{j|i} + v_{i|j}\right) - v_{j|i}v_{i|j}$$

The distances for the reduced space are estimated by:

$$w_{ij} = \left(1 + a\big(d(x_i, x_j)_2^{2b}\big)\right)^{-1}$$

Where a and b are hyperparameters which default to approximately 1.93 and 0.79 respectively, when hyperparameters *min_dist* = 0.1 and *spread* = 1. The hyperparameters a and b are determined by a non-linear least square fit based on the *min_dist* and *spread* hyperparameters that control the tightness of the clusters. If a and b are set to 1, a Student t-distribution is used, such as in t-SNE. This was not applied in this work.

2. Cost-function optimization

To seek for the most proximate structure to the original data, a cost function optimization procedure is applied in a second stage. UMAP uses binary cross-entropy cost function instead of the *KL-divergence* used in t-SNE. The reduced topological layout that optimizes the cross-entropy loss between both topological structures is computed as follows:

$$C = \sum_{ij}\left[v_{ij}log\frac{v_{ij}}{w_{ij}} + (1 - v_{ij})log\frac{1 - v_{ij}}{1 - w_{ij}}\right]$$

Where $v_{ij}log\frac{v_{ij}}{w_{ij}}$ is the data points attractiveness (similarity) and $(1 - v_{ij})log\frac{1 - v_{ij}}{1 - w_{ij}}$ is the data points repulsiveness (dissimilarity), v_{ij} is the simplex low-dimensional weight (data points weight in high dimensions), and w_{ij} is the weight of data points in low-dimensional distribution, which are ultimately expressed in terms of probability. The second term in this function enables this algorithm to capture global data structure, contrarily to the t-SNE method.

Summarily, the basic methods' properties are listed in the following Table 1.

Method	Type	Computational Complexity	Non-linear	Manifold Learning
PCA	Deterministic	$\mathbb{O}(max(n^2p,\ np^2))$		
SVD	Deterministic	$\mathbb{O}(dcn)$		
RP	Stochastic	$\mathbb{O}(ckn)$		
CMDS	Deterministic	$\mathbb{O}(n^3)$	✓	
NMDS	Deterministic	$\mathbb{O}(n^2)$	✓	
IM	Deterministic	$\mathbb{O}(n^2(p + log\ n))$	✓	✓
LLE	Deterministic	$\mathbb{O}(dn^2) + \mathbb{O}(dnk^3) + \mathbb{O}(dn^2)$	✓	✓
Barnes-Hut t-SNE	Stochastic	$\mathbb{O}(nlog(n))$	✓	✓
UMAP	Stochastic	$\mathbb{O}(Kn)$	✓	✓

Table 1 - Basic DR methods' properties. Linear or non-linear; computational complexity in terms of: n - the number of observations, p - the number of features in the original data, k - the selected number of nearest neighbours, c – number of nonzero entries, d - number of dimensions; K - number of edges on fuzzy graph; Use of manifold learning.

2.2.6 Hierarchical Clustering

Hierarchical clustering is commonly used in transcriptome studies [65]. Even though it scales quadratically in both time and memory requisites, current machines are normally able to perform it in a relatively fluent way, just needing slightly more memory. It merges each sample or cell data into larger clusters or splits larger clusters into smaller sets consecutively, in an agglomerative or divisive approach, respectively [65]. We have adopted the former procedure for this study, as it is more broadly applied. Primarily, all the data points are considered as clusters. Based on a matrix of pairwise similarities between each cluster, it subsequently merges data points into a new cluster. This process is reiterated until all data points are comprised in a single cluster. The definition of similarity between two clusters may consist in the average similarity between all pairs of data points in different clusters, which is known as average linkage. This was the definition used for this study, as it is generally considered to work better with standardized transcriptome data. This method generates a dendrogram to visualize the merged clusters in a structure representation with increased detail in relation to other clustering methods.

2.2.7 K-means Clustering

From the various clustering methods available, k-means clustering [66] is the most widely used in transcriptome studies and in other fields [67], [68] as it is straightforward and effective. It is a key element of frequently used scRNA-Seq clustering methods [69] [70] and quantification tools [71]. This technique splits the data into k clusters by allocating data vectors to the nearest cluster centroid, which is firstly determined. These centroid positions are then iteratively optimized.

Formally, k-means algorithm is based on the concept of grouping n points into k sets ($\leq n$ sets), $\{S_1, ..., S_k\}$. It aims to minimize the intra-cluster variance by minimizing the squared deviation of the data points within each cluster, also known as within-cluster sum of squares (WCSS), with the following objective function:

$$Cost(S, M) = \sum_{i=1}^{k} \sum_{m_j \in S_i} \left\| m_j - \mu_i \right\|_2^2$$

Where μ_i is the mean of the vectors in S_i and M is a data matrix comprising the data vectors as rows. The square deviation is the sum of squared distances between the data vectors and the centroids of their assigned clusters. In this thesis we used the Euclidean distance for this method, as it is its most appropriate distance metric, because the sum of squared deviations from centroid is equal to the sum of pairwise squared Euclidean distances divided by the number of points.

To compute the optimization procedure, we mostly used *Hartigan-Wong* algorithm [72] for the GTEx dataset, which is generally known to operate the best among the existing algorithms. However, this algorithm did not converge properly for the various reduced spaces produced by the different feature extraction methods on the scRNA-seq dataset and for the LLE method on the bulk RNA-Seq dataset. This is due to the presence of extremely similar values for some data points. In alternative, we applied *MacQueen's* algorithm [73] for both, as it is a good alternative for the described situation. This latter approach uses the total sum of squares rather than the within sum of squares used in the Hartigan-Wong algorithm. With this substitution, the algorithm converged adequately in all the runs.

Chapter 3

Tuning Hyperparameters

3.1 General approach and distance measure applied

Several feature extraction methods, especially non-linear methods, require the fine tuning of their hyperparameters. This task is particularly important, but it is often overlooked. Comparison of the methods with the default hyperparameters may not be the most appropriate approach. Fine-tuning can return very different and much improved outcomes. This procedure was performed *ad hoc* for each method, in association with guidelines in literature. This way, hyperparameters were selected according to the best resultant representation and to the existent recommendations. For CMDS, NMDS, Isomap and HC clustering, a distance metric must be selected. We used a correlation-based distance measure which is typically applied to gene expression datasets to identify clusters of samples with similar genetic expression profiles. This metric assumes two observations or experiments to be similar based on the correlation of the respective features. This way, two samples that are perfectly correlated by their genes are considered to have zero distance between. Thus, this measure is produced by subtracting the correlation coefficient from 1:

$$dist(x_i, y_i) = 1 - cor(x_i, y_i)$$

Where *cor* can be measured as the Spearman or Pearson correlation [74]. Pearson correlation may be defined as the division of the pairwise covariance of variables by the product of their standard deviations. In contrast, Spearman is a rank-based metric that evaluates the concordance between the ranks of the raw values. In this extent, Spearman correlation coefficient is less sensitive in comparison, although more adequate in the presence of outliers. Here, Pearson coefficient was selected for the construction of the dissimilarity matrix applied in CMDS, NMDS, and HC techniques. This coefficient can be computed as:

$$d_{cor}(x, y) = 1 - \frac{\sum_{i=1}^{n}(x_i - \bar{x})(y_i - \bar{y})}{\sqrt{\sum_{i=1}^{n}(x_i - \bar{x}) \sum_{i=1}^{n}(y_i - \bar{y})^2}}$$

It should be remarked that, in this case, the outcome obtained by Pearson correlation measures and Euclidean distances are comparable, as there is a functional relation between both of these metrics when dealing with standardized data, very close to 1. This relation is defined as follows:

$$d_{euc}(x,y) = \sqrt{2m[1 - r(x,y)]}$$

3.2 Specific adopted approach

Commonly, for a better adjustment of the hyperparameters for a certain feature extraction method, a *Silhouette heuristic* strategy may be adopted over a grid of possible hyperparameter options and the configuration that optimizes it is selected [30] [75] [76]. This heuristic is uniquely based on the silhouette coefficient metric: after k-means implementation on the reduced space, the Silhouette coefficient metric, which is a metric which estimates the validity of a certain distribution of data vectors corresponding to a possible clustering of distinct cell or tissue type, is computed regarding the classes assigned by this clustering method. After its implementation, we verified that this heuristic corroborated the qualitative evaluation, *i.e.*, the configurations that returned the best silhouette score corresponded to the visualizations that, visually and subjectively, presented a better result. As previously described, this qualitative evaluation was based on a correct segregation by class and cluster compactness. However, we did not verify this relation on the clustering evaluation performance. In fact, after taking all the classification and clustering validity evaluation metrics in consideration, we noticed that there was not always a positive relation between the silhouette coefficient metric and the best possible outcomes.

Therefore, in the pursuance of the best possible representation, we did not adopt this heuristic optimization, but rather a more comprehensive one, taking all the evaluation metrics presented into consideration. Even though this approach requires much more time for the implementation and obtainment of each method reduced representation, we think that it is much more adequate, especially for a proper quantitative evaluation and comparison between the various methods' representations.

3.3 Relevant hyperparameter considerations for each method

Some of the studied methods present important particularities to take into consideration for the estimation of the most appropriate hyperparameters. Particularly, t-SNE, LLE, Isomap and UMAP are the ones which required an optimization step. To choose the number of neighbours for the Isomap algorithm, multiple runs with a broad range of values were performed. For the LLE method, we confirmed that the number of neighbours' parameter has an even more notorious impact, as different values can yield very different results. Various extensions of LLE were presented in recent years, and the majority addressed this issue. However, none of these algorithms overcome all the specific drawbacks of this method in respect to this parameter. Therefore, there are no sufficiently consistent procedures to apply yet. As a linear space of $k - 1$ dimensions at most constitutes the span of a set of k nearest neighbours for this algorithm, the number of nearest neighbours should be higher than the reduced space dimensionality. Yet, selecting a low number of nearest neighbours may raise several problems, because the eigenvector problem of finding the embedding data points becomes singular [77].

For t-SNE method, the most relevant parameter to be computed is the perplexity value. Guidelines for the selection of this parameter are scarce and do not have a solid mathematical basis. The most consensual rule of thumb for this parameter in the case of scRNA-Seq data is to consider 1% of the sample size [78]. This consideration corresponded to perplexity 49 for our single-cell dataset, which did not yield the best visualization and k-means clustering evaluation results. The selected value for our dataset was perplexity 70, as it demonstrated the best qualitative and quantitative evaluation. This value is much higher than the default perplexity 30, which also did not return a good result. Giving an example, to tune this method, the values analysed ranged from 20 to 80, with intervals of 10. Once the value 70 was detected as showing the best results among these values, a range of 60 to 80 was also verified with a unit interval. In addition, we increased the learning rate to $n/12$, where n is the number of samples or cells, as recommended in [79]. A PCA initialization was also adopted for this method. This initialization permits a fast, deterministic, and reproducible implementation. This method was executed with the *Rtsne* function from *Rtsne* package which uses a Barnes-Hut C ++ implementation on the original t-SNE algorithm. For this reason, the implementation of this method is considerably fast in R programming language. In contrast to what may be presumed, finding the optimal perplexity value based on the minimum of Kullback-Leibler (KL) divergence is not an adequate approach. Even though t-SNE is based on its optimization, it is generally known that this divergence decreases monotonically with higher perplexity values, which would then bias the indication of the perplexity value to apply. The UMAP algorithm has also various parameters to tune, and as a recent method, there are not many consistent and reliable guidelines yet. This way, it is also important to experiment with different combinations of parameters to select the best configuration. A PCA initialization step was also implemented in this technique, for the previously stated reasons.

Chapter 4

K-means Clustering-based Evaluation

4.1   Framework

Several visualizations were generated for the different methods applied and a qualitative comparison of the results was made. However, this is a subjective evaluation. For a more precise evaluation, we implemented a k-means clustering-based quantitative methodology. This algorithm was performed on the reduced space of every method for dimensions from two to four, and on the full datasets without any DR, to establish the ground truth class labels and performance. As we already had the information that there were 53 types of tissues in GTEx dataset and 10 cell types in scRNAseq dataset, we set this number of clusters (k) for the algorithm to generate and executed it multiple times (100), selecting the clustering result with the higher silhouette coefficient score in order to optimize its configuration.

To compare the performance of the various DR methods, a contingency table was generated. This table contained the frequency of samples for each tissue type (or cell type, in the case of scRNA-Seq data) in the respective allocated cluster. The specific cluster that comprised more samples for each was considered the true reference cluster of that tissue type for the subsequent statistical analysis.

Let's consider a Toy example, represented in Figure 6. Suppose that we have three tissue types in a dataset – T_1, T_2 and T_3. Their distribution is, for instance, the one represented in Figure 6 (a). For k-means clustering, we set k as the number of tissue types, *i.e.*, $k = 3$, obtaining three clusters – K_1, K_2 and K_3. Then, we perform a mapping based on the cluster that comprised more samples for each tissue type. The resultant table is represented in Figure 6 (b), with cluster T_1 being allocated to cluster K_1, T_2 to K_2, and T_3 to K_1, as well. Following this ground truth labels mapping, we then calculated the contingency table for each tissue type (Figure 6 (c)).

Finally, we aggregate these results and obtain a final contingency table with the number of true positives (30+15+50=95); false negatives (20+5+50=75); false positives (50+50+30=130); and true negatives (100+100+40=240), as represented in Figure 6 (d).

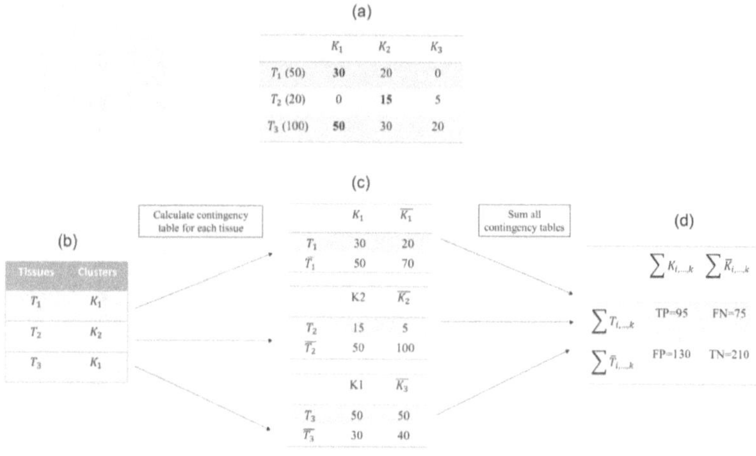

Figure 6 - Schematic representation of the specific methodology of contingency table applied. It is first necessary to sum the total counts of samples of each tissue (T_1, T_2 and T_3) allocated to a cluster (K_1, K_2 and K_3). A contingency table is then calculated for e each tissue based on the reference of the most frequent cluster for each as the true cluster. Finally, all the contingency tables are summed, and a final contingency table is obtained for each DR method clustering result.

This generated table is used to calculate the classification statistical metrics for each DR clustering result. These metrics were, namely, overall Accuracy, Precision, Recall, Specificity, F1-score, and Cohen's kappa.

Furthermore, for a more precise and specific performance measurement, several clustering validity measures were also applied, namely the Adjusted Rand index, V-measure, Homogeneity, Completeness, Fowlkes–Mallows score, and the mentioned Silhouette coefficient. These measures compare the vector of the predicted clusters with the one containing the ground truth class labels, except for Silhouette coefficient, which measures the consistency within each cluster. All these applied metrics will be overviewed in the next sections.

The full k-means clustering-based evaluation framework is represented in Figure 7.

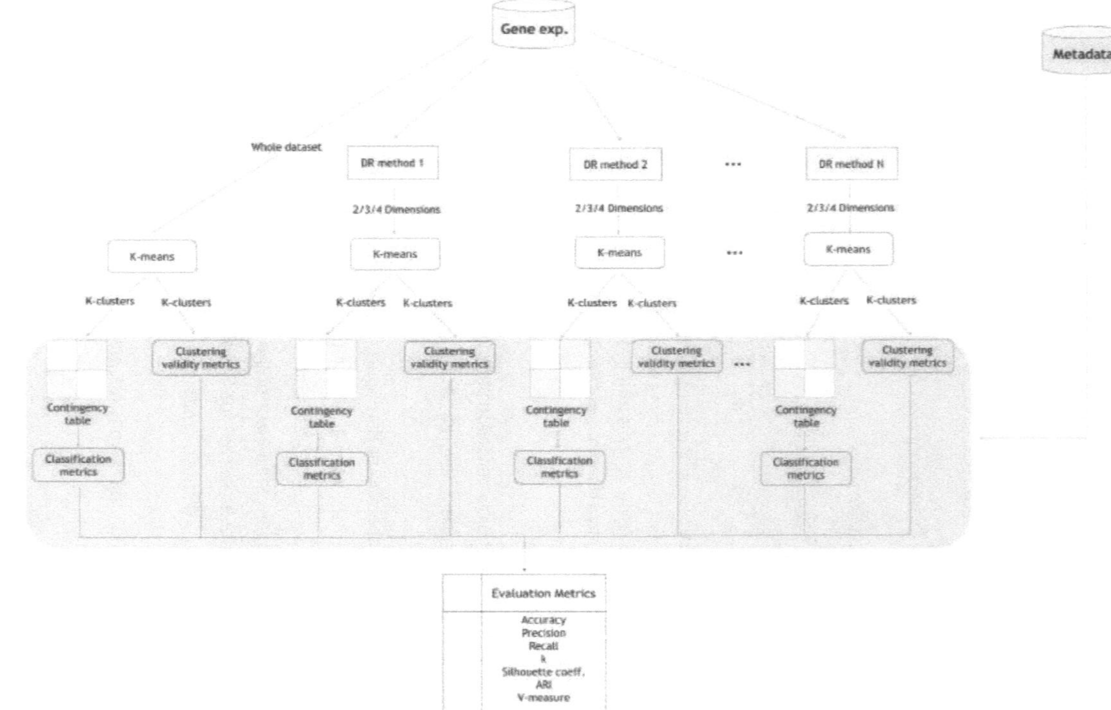

Figure 7 - K-means clustering-based evaluation scheme. K-means was performed on the full dataset and on the reduced spaces from two to four dimensions. We set the number of clusters to generate as the number of tissues or cells in each dataset (k) and executed it 100 times, selecting the clustering result with the higher silhouette coefficient score. Then, the categories of tissue/cell type information were mapped to the clustering result, and the classification and clustering validity metrics were applied. To generate classification metrics, a specific contingency table methodology was created, as clustering is an unsupervised method. For each representation, the cluster that comprised more samples was considered the true reference cluster for each tissue. Then, all the contingency tables of each tissue/cell type were summed to produce a final contingency table for each method. Once the best configuration was selected for each, a final table with every metric for all the methods was retrieved.

4.2   Classification Metrics

In this section, the classification metrics applied for the k-means clustering-based evaluation are overviewed. All the metrics formulas are represented in Table 2.

4.2.1 Accuracy

In an initial stage, a binary classification problem is considered in this work, as it is evaluated for each tissue/cell type if the sample is, or is not, allocated to its most frequent cluster. For this problem type, the clustering accuracy, which is the ratio of correct predictions to the total number of samples, can be calculated by taking the average of the values lying across the "main diagonal" of the confusion matrix. After measuring this metric for all the tissue types, and considering that both datasets are class-imbalanced, an average of this metric was taken.

4.2.2 Precision

Precision quantifies the proportion of correctly classified instances among all positive predictions made. For imbalanced classification problems and with more than two classes, as it is the case for the data analysed in this thesis, precision is calculated as the sum of true positives across all classes divided by the sum of true positives and false positives across all classes.

4.2.3. Recall

Recall, also known as sensitivity, quantifies the proportion of actual positive predictions that were identified correctly, providing an indication of missed positive predictions, unlike the precision measure. For imbalanced data, recall is calculated as the sum of true positives across all classes divided by the sum of true positives and false negatives across all classes.

4.2.4 F-score

The F-score is the harmonic mean of the precision and recall, combining both metrics into a single score that captures their both criteria. Giving equal weight to both precision and recall, it is the variant most frequently applied on imbalanced data.

4.2.5 Cohen's Kappa

Cohen's kappa is one of the most used quantitative statistics to measure the level of agreement between two raters for categorical items. It adjusts these values for concordance that could be expected due to chance only. This statistic ranges from 0 to 1, in which 0 corresponds to an agreement equivalent to chance; 0.1-0.2 to a slight agreement; 0.21 – 0.40 to a fair agreement; 0.41 – 0.60 to a moderate agreement; 0.81 – 0.99 to a near perfect agreement and 1 to a perfect agreement.

4.3 Clustering validity metrics

Generally, for a cluster analysis to be considered of high quality, it should present high within-cluster similarity and a low between-cluster similarity [80]. This means that it should aim for homogenous clusters and distinct to other clusters. In this section, the clustering validity metrics used in this project to evaluate the k-means clustering of each projection are addressed. All the metrics formulas are represented in Table 2.

4.3.1 Adjusted Rand Index

The Adjusted Rand Index is a commonly used measure to evaluate the ability of a clustering algorithm to match the known classes. It measures the similarity between the clustering obtained with the ground truth labels by using a contingency table. With this table, it assesses all pairs of samples and counts the ones that are allocated in the same cluster, estimating an index. The expected index imputable to a particular cell in the contingency table is defined by the fraction of pairs of labels either present in the same group or in a different group. This fraction is the amount of pairs in the row multiplied by the amount of pairs in the column divided by the total amount of pairs, while the maximum number of object pairs is given by the average number of possible pairs in the clustering and the true labels [81].

4.3.2 Homogeneity

To be considered a perfectly homogeneous clustering, each unique cluster must have data points which are members of a unique class label. Thus, this measure scores higher for fewer distinct classes present in each cluster [82].

4.3.3 Completeness

In a symmetrical way to homogeneity, completeness evaluates the clustering by the amount of data points that pertain to each class which are allocated for a single cluster. Expanding the homogeneity of a clustering pattern frequently diminishes its completeness, and vice-versa [82]. This evaluation can be computed by the estimation of the conditional entropy of the projected cluster distribution provided the class of the datapoint, $H(K|C)$. In the case of perfect completeness, $H(K|C) = 0$ and the completeness score is 1.

4.3.4 V-measure

V-measure is the harmonic mean of the homogeneity (h) and completeness (c) scores of the clustering, just as F-measure is for precision and recall [83], measuring how successful these criteria were met.

4.3.5 Silhouette coefficient

Silhouette coefficient is commonly used to evaluate the performance of embedded methods [84][85]. It quantifies the similarity of a sample or cell to its own cluster in comparison to others. In this way, this metric measures how near are a cell or sample from the same cell/tissue type compared to the cells or samples from other types. It uses a distance metric to calculate the dissimilarity of that data point to all other data points in the same cluster (mean intra-cluster distance, a) and the distance between a data point and the nearest cluster that does not include that data point (mean nearest-cluster distance, b). These distances a and b are then averaged for each sample [86]:

$$silhouette = \frac{b(x) - a(x)}{\max\left(a(x), b(x)\right)}$$

Where $a(x)$ represents the average Euclidean distance between x and the other cells or tissues of the same type, and $b(x)$ is the average Euclidean distance between x and cells or tissues in the nearest distinct class. Thus, this metric evaluates the performed clustering by examining the compactness within clusters and a correct partition among data points.

4.3.6 Fowlkes-Mallow Score

Fowlkes-Mallow score is computed as the geometric mean between precision and recall classification metrics. In this case, the number of true positives was measured by the number of pairwise points that pertained to the same cluster in both the constructed ground truth labels vector and the clustering result labels; the number of false positives by the number of pairwise points that pertained in the same clusters in truth labels but not in the clustering labels; and the number of false negatives by the number of pairwise points that pertained in the same clusters in clustering labels but not in the truth labels.

Metric	Formula
Average Accuracy	$\frac{\sum_{i=1}^{c} \frac{tp_i + tn_i}{tp_i + tn_i + fp_i + fn_i}}{c}$
Precision	$\frac{\sum_{i=1}^{c} \frac{tp_i}{tp_i + fp_i}}{c}$
Recall	$\frac{\sum_{i=1}^{c} \frac{tp_i}{tp_i + fn_i}}{c}$
F-score	$\frac{2 \cdot (Precision \cdot Recall)}{Precision \cdot Recall}$
Cohen's Kappa	$k = \frac{p_o - p_e}{1 - p_e} = 1 - \frac{1 - p_o}{1 - p_e}$
Adjusted Rand Index	$\frac{\sum_{ij}\binom{n_{ij}}{2} - \left[\sum_i \binom{a_i}{2} \sum_{ij} \binom{b_j}{2}\right] / \binom{n}{2}}{\frac{1}{2}\left[\left[\sum_i \binom{a_i}{2} + \sum_{ij} \binom{b_j}{2}\right]\right] - \left[\sum_i \binom{a_i}{2} \sum_{ij} \binom{b_j}{2}\right] / \binom{n}{2}}$ where $ai = \Sigma j\, ni$ and $bj = \Sigma i\, nj$.
Homogeneity	$\begin{cases} 1 & \text{if } H(C,K) = 0 \\ 1 - \frac{H(C\mid K)}{H(C)} & \text{else} \end{cases}$ where $H(C\mid K) = -\sum_{k=1}^{\lvert K\rvert}\sum_{c=1}^{\lvert C\rvert} \frac{a_{ck}}{N} \log \frac{a_{ck}}{\sum_{c=1}^{\lvert C\rvert} a_{ck}}$ $H(C) = -\sum_{c=1}^{\lvert C\rvert} \frac{\sum_{k=1}^{\lvert K\rvert} a_{ck}}{n} \log \frac{\sum_{k=1}^{\lvert K\rvert} a_{ck}}{n}$
Completeness	$\begin{cases} 1 & \text{if } H(K,C) = 0 \\ 1 - \frac{H(K\mid C)}{H(K)} & \text{else} \end{cases}$ where $H(K\mid C) = -\sum_{c=1}^{\lvert C\rvert}\sum_{k=1}^{\lvert K\rvert} \frac{a_{ck}}{N} \log \frac{a_{ck}}{\sum_{k=1}^{\lvert K\rvert} a_{ck}}$ $H(K) = -\sum_{k=1}^{\lvert K\rvert} \frac{\sum_{c=1}^{\lvert C\rvert} a_{ck}}{n} \log \frac{\sum_{c=1}^{\lvert C\rvert} a_{ck}}{n}$
V-measure	$V_\beta = \frac{(1+\beta)h \cdot c}{\beta \cdot h + c}$
Silhouette Coefficient	$\frac{b - a}{\max(a - b)}$
Fowlkes-Mallow Score	$\sqrt{\frac{tp}{tp+fp} \cdot \frac{tp}{tp+fn}}$

Table 2 - K-means clustering-based evaluation metrics formulas. The formulas for each metric used in this project are represented, where: tp – true positives; tn – true negatives; fp – false positives; fn – false negatives; p_o - the comparative agreement between raters; p_e - the theoretical probability of chance agreement; Suppose that the given dataset is partitioned as c subsets $\mathrm{C} = \{\mathrm{C}_1, \dots, \mathrm{C}_c\}$, the estimated clusters are composed of s subsets, $\mathrm{K} = \{\mathrm{K}_1, \dots, \mathrm{K}_s\}$, and a contingency table $[n_{ij}]$ is used where $n_{ij} = \lvert \mathrm{C}_i \cap \mathrm{C}_j \rvert$; a_{ck} is the number of data points that are members of class C_i and elements of cluster K_i, i.e. $a_{ck} = \lvert \mathrm{C} \cap \mathrm{K} \rvert$; N – dataset,; h – homogeneity; c – completeness; $\beta\ (> 0)$ - can be used to attribute more weight to either metric; a - mean intra-cluster distance; b – mean nearest-cluster distance.

Chapter 5

GTEX Dataset

5.1 Data source and pre-processing

The Genotype-Tissue Expression (GTEx) consortium RNA-Seq gene expression data (v6p) was used for this analysis. The dataset comprises samples from various types of healthy tissues from various individuals. It contains expression values for 56238 genes, including the coding and non-coding ones across 8555 different tissue samples. Samples are divided into 31 types of tissues, further subdivided into 53 types. Assessments were made based on the 53 tissue types subdivision. The number of samples per tissue is unevenly distributed, and it varies between 5 as the minimum number of samples to 430 as the maximum, the median value being 119 (Supplementary Figure 1).

Primarily, a data pre-processing step was applied. Given the very high dimensionality of the dataset, we started by selecting the most informative features and discard those that in principle are expected to bring little information to the analysis process. In particular, these will be genes with low expression and variability. Moreover, certain gene biotypes have a noisier expression profile, such as small RNAs or non-coding RNAs. Therefore, we started by selecting the most relevant genes by filtering protein-coding genes and long-intergenic noncoding RNA (lincRNAs) genes, as these are the ones which are more significantly expressed and informative. The resulting number of genes for this dataset after this filtering process was 26872 genes.

A logarithmic transformation and a verification for missing data or outliers was also performed. After verification, this dataset did not present any missing data or outliers. Sample annotation data was used to produce the labels of the tissue type in detail correspondent to each sample. The cumulative variance scree plot of the variance explained for this data by PCA is represented in Supplementary Figure 13. In this plot, it is illustrated that only 27 PCs are required to explain 85% of the data variance. The regular scree plot of variance explained, generally used to determine the number of PCs to use for the data analysis, shows an 'elbow' between PCs 6 and 12. The first 6 PCs were considered for the analysis.

5.2 Results

In this section, the qualitative results on the GTEx dataset of the various visualizations generated for all the DR methods implemented is presented, as well as the respective quantitative results of the k-means clustering-based evaluation.

5.2.1 Qualitative evaluation results

Figure 8 depicts the visualization results of the various DR methods in the first two dimensions. Each dot represents a sample, and each colour a specific tissue type. Generally, we observe that these visualizations demonstrate a noticeable segregation by tissue type. Some tissues such as brain, whole blood, testis, and pituitary appear to be more easily separated than others. Indeed, it is known that these tissues have a more distinguishable genetic expression. We may also observe that some methods distinctly group samples of the same tissue in a clearer way. Nonlinear DR methods, in general, seem to depict this tendency in relation to linear methods. The visualization results for the remaining axis are consistent with these results, and are provided in Supplemental Figure 2, 3, 4, 5 and 6. It is also noticeable that some samples do not aggregate with their tissue type, which may occur for various reasons, such as random noise, underlying medical conditions of the patients, mislabelling or contamination.

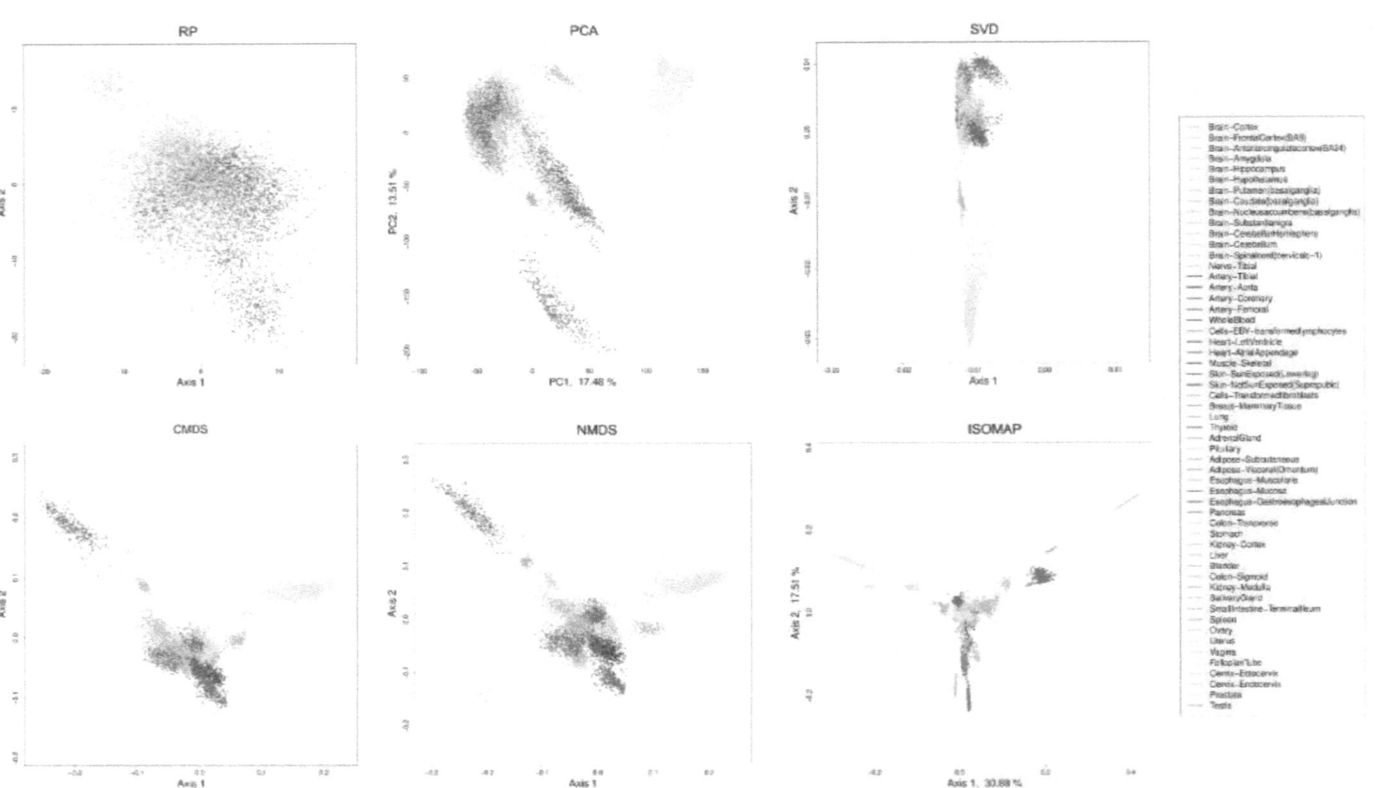
RP
PCA
SVD
CMDS
NMDS
ISOMAP
Axis 1
Axis 2
PC1, 17.48 %
PC2, 13.51 %
Axis 1, 30.88 %
Axis 2, 17.51 %
Brain-Cortex
Brain-FrontalCortex(BA9)
Brain-AnteriorcingulatecortexBA24)
Brain-Amygdala
Brain-Hippocampus
Brain-Hypothalamus
Brain-Putamen(basalganglia)
Brain-Caudate(basalganglia)
Brain-Nucleusaccumbens(basalganglia)
Brain-Substantianigra
Brain-CerebellarHemisphere
Brain-Cerebellum
Brain-Spinalcord(cervicalc-1)
Nerve-Tibial
Artery-Tibial
Artery-Aorta
Artery-Coronary
Artery-Femoral
WholeBlood
Cells-EBV-transformedlymphocytes
Heart-LeftVentricle
Heart-AtrialAppendage
Muscle-Skeletal
Skin-SunExposed(Lowerleg)
Skin-NotSunExposed(Suprapubic)
Cells-Transformedfibroblasts
Breast-MammaryTissue
Lung
Thyroid
AdrenalGland
Pituitary
Adipose-Subcutaneous
Adipose-Visceral(Omentum)
Esophagus-Muscularis
Esophagus-Mucosa
Esophagus-GastroesophagealJunction
Pancreas
Colon-Transverse
Stomach
Kidney-Cortex
Liver
Bladder
Colon-Sigmoid
Kidney-Medulla
SalivaryGland
SmallIntestine-TerminalIleum
Spleen
Ovary
Uterus
Vagina
FallopianTube
Cervix-Ectocervix
Cervix-Endocervix
Prostate
Testis

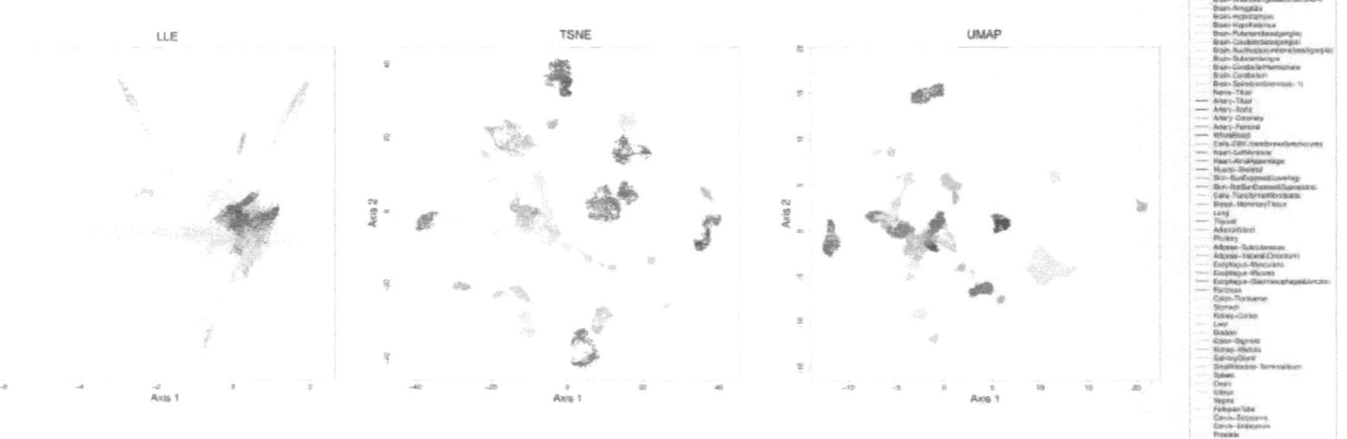

Figure 8 - Visualization results in the first two dimensions for all the DR methods performed on the GTEx dataset. A clear segregation by tissue type can be seen for the various methods, even though UMAP and t-SNE show clear patterns.

Among linear methods, the difference between PCA and SVD visualization is clear, as in PCA visualization the different tissue types are much more segregated and distinguishable. Notably, we can observe that the first PC explains 17,48% of the data, and the second PC explains 13,51% of the data, in a total of about 31% of data variance explained solely by the first two PCs. We observe that the first PC divides brain tissues, and pituitary and testis tissue from the remaining tissues, and the second PC divides the whole blood from the remaining tissues. Such division does not occur for the third PC, but for the fourth it does, although less prominently. The third PC separates the skeletal muscle, heart, artery, genito-urinary and adipose tissues from lung, thyroid, whole blood, brain, skin, colon, pituitary, testis, adrenal gland, and stomach tissues. The fourth PC mainly separates the whole blood and cells of EBV-transformed lymphocytes and lung from the other tissues, in which skin, skeletal muscle, artery aorta and oesophagus stand out as being more separated from the other types. The RP method visualization seems to represent a much inferior capability of distinguishing the different tissue types, although presenting some separation for muscle-skeletal, genito-urinary and testis samples.

For non-linear methods, one can say that CMDS and NMDS aggregates the samples from the same tissue in a more compact way than linear methods such as PCA and SVD. Yet, the same does not apply when comparing CMDS and NMDS to the other non-linear methods. This discrepancy probably relates to the fact that data lies along a nonlinear embedding, and multidimensional scaling is not able to "unwrap" this type of topology and uncover its fundamental relations as the other non-linear DR methods do. Comparing CMDS to NMDS visualizations outcomes, a better segregation for some tissue types such as genito-urinary/gastrointestinal systems and cells of EBV-transformed lymphocytes is observed for the NMDS visualization. This result may also unveil a better adequacy of NMDS method over CMDS for this type of data, probably related to its ranking ordination procedure. We can observe the best outcomes in distinguishing tissue types for both t-SNE and UMAP techniques. Visually, Isomap and LLE seem to achieve a reasonable separation between tissue types, although not as good as t-SNE and UMAP. Yet, impressively, the first axis of the Isomap method explains 30.88% of the data variance, and the second axis explain 17,51%, in a total of 48,39% of explained data variance with only the first two axis of this feature extraction method. In contrast with other methods, t-SNE and UMAP visualizations do not exhibit data points in transition between clusters. Indeed, these methods do not focus on these transitions, as their main aim is to identify the clusters of similar samples present in the data. However, UMAP agglomerates are more visually identifiable than t-SNE, as t-SNE structures are more scattered in space. Additionally, UMAP presents the advantage that insights can be taken based on the cluster proximity, as it is efficient at preserving both local and global structure of data, contrarily to t-SNE method.

Observing the 3D plots of the DR method most used for this type of data (PCA), and the most effective in terms of visualization (t-SNE and UMAP), the segregation by tissue type is even more prominent (Figure 9). The difference between the PCA representation and t-SNE and UMAP representations is, once again, clearly visible. Importantly, the cluster compactness observed for the UMAP representation in the two-dimensional plot is even more evident in these three-dimensional plots.

Hierarchical clustering method visualization clearly represents a cluster separation for each tissue type as well (Figure 10). The tissues that seem to be more distant from others are testis and whole blood tissues, an observation that is also verified in the previously mentioned DR methods' visualizations. Importantly, it can also be observed that some tissue types present subclusters within its tissue cluster, which mostly correspond to a specific part of the tissue in detail, as can be seen in the colour legend. This is especially the case for the brain tissue and for the transformed fibroblasts. Reversely, there is also aggregation of some tissue types with others. This pattern can be more notoriously seen for some tissue types such as: pituitary and brain tissue; artery aorta and skeletal muscle; skin of lower leg, skin of suprapubic and oesophagus muscle; and colon transverse and oesophagus.

Nevertheless, as visualization outcomes are not appropriate to unequivocally evaluate the quality of DR outcomes, we proceeded to the quantitative evaluation in the next section by examining the performance of k-means clustering in the reduced subspaces.

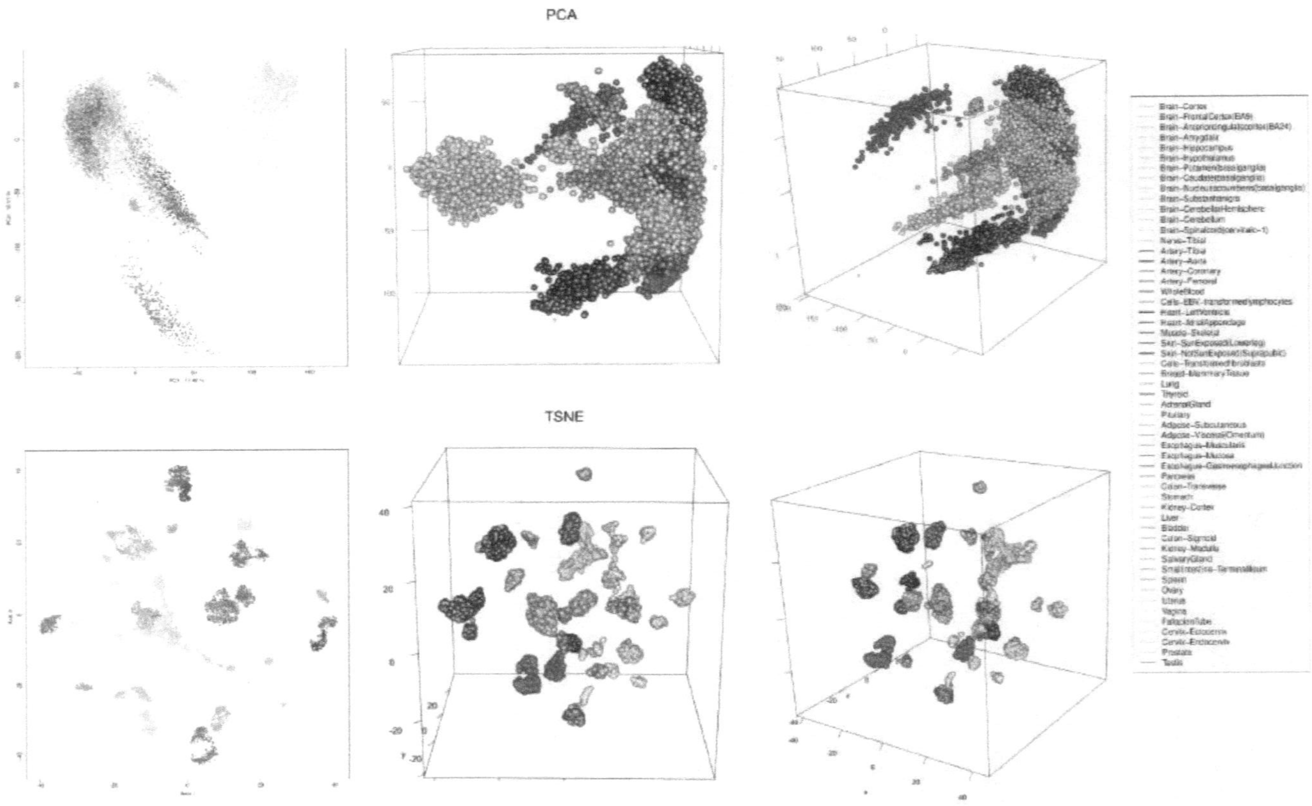

PCA

TSNE

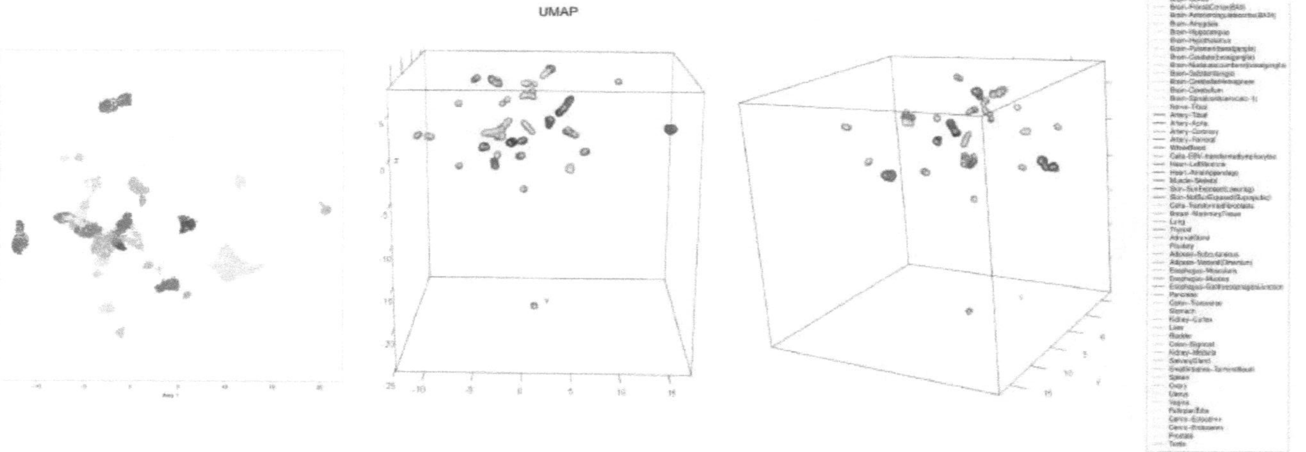

Figure 9 - Visualizations of PCA, t-SNE and UMAP representations on the GTEx dataset in the first two dimensions (left) and in three dimensions, on two different angles (centre and right). A segregation by tissue type is even more prominent in this 3D space, as well as the difference between the PCA representation and t-SNE and UMAP representations. Samples are coloured by their tissue type.

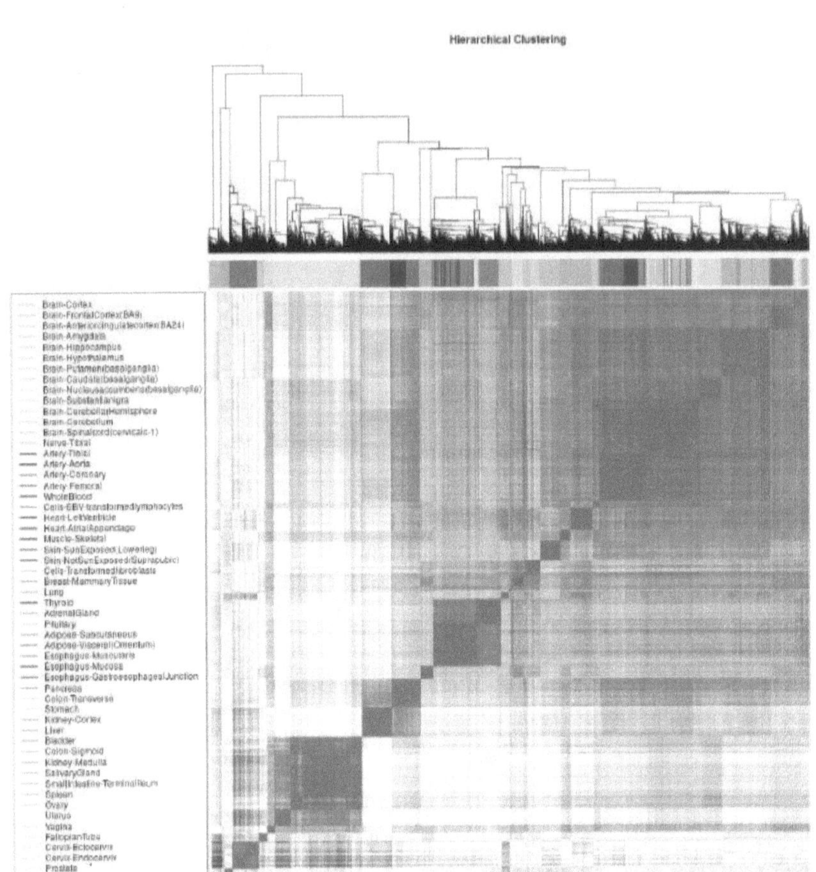

Figure 10 - Hierarchical Clustering visualization result for the GTEx dataset. A segregation by tissue can clearly be visualized, as well as some subclusters within some tissues, generally corresponding to a specific part of the tissue in detail. Correlation between samples is represented by a colour code from blue (negative correlation) to red (positive correlation). White cells represent that there is no correlation between samples.

5.2.2 Quantitative evaluation results

Quantitative evaluation based on k-means clustering-based (Figure 11) shows that non-linear manifold learning methods have a better performance. Particularly, UMAP, t-SNE and LLE demonstrate a superior performance, both for classification metrics and for clustering validation metrics. Notably, we also verify that, up to four dimensions, better results are still obtained with the reduction performed by these methods than when the full dataset is used. This is even observed with only three dimensions for t-SNE and UMAP. In accordance with the qualitative results, we observe that CMDS and NMDS were the non-linear methods which showed the worst quantitative evaluation scores. These results once again probably relate to the fact that data lies along a nonlinear embedding, and Multidimensional scaling does not perform so well with this data topology. Among linear methods, PCA and SVD show a considerable advantage over RP method. Up to three dimensions, PCA shows generally better scores than RP and SVD, and even over non-linear methods such as CMDS and NMDS. However, in the four-dimensional space it is surpassed in most metrics by SVD, CMDS, and NMDS. Equivalently to qualitative results, RP method seems once again to be the one that indicates the poorest performance.

In Figure 12, tissues frequency distribution by each cluster for the full dataset (baseline) and for each DR method are shown. Tissues are ranked in a descendent order based on the true positive rate (recall metric). This way, the tissues that are on top of this figure are the ones which are most frequently correctly classified. We can generally observe better results for non-linear dimensionality methods, once again corroborating the previous evaluation analyses. UMAP, t-SNE and LLE also stand out in these results as the ones demonstrating a higher specificity of each tissue for the respective top cluster. This constitutes another indicator of better projections for these methods, with a better approximation to the full dataset information. By comparing to the full dataset heatmap, this observation is reinforced. For a plainer differentiation of these results, we can see the two-dimensional heatmap illustrated in Supplementary Figure 7. From these heatmaps it can be more evidently noted that, notably, LLE is the method that presents more tissue type specificity, followed by UMAP and t-SNE. The amount of perfectly assigned tissue types is about 60% for LLE, while for t-SNE this amount is about 23% and for UMAP is about 30%. Even though Isomap falls evidently behind these methods, it has slightly better results than NMDS and CMDS. Considering linear methods, PCA is, clearly, the one which presents more tissue specificity. The correspondent heatmaps results in three and four dimensions (Supplementary Figure 9 and 11) also demonstrate generally accordant results in relation to the overall clustering metrics results.

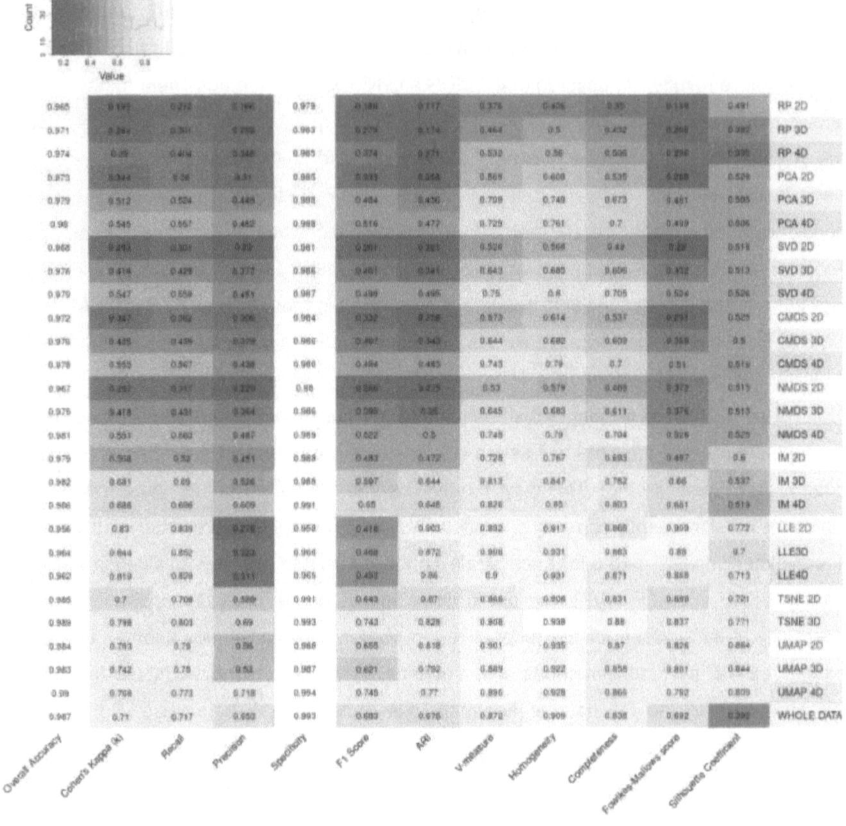

Figure 11 - Heatmap table of the k-means clustering evaluation metrics results. Values range from 0 to 1. A notable improvement from the linear methods to the non-linear methods can be clearly observed, as well as a better performance of t-SNE and UMAP methods in in three and four-dimensional data, respectively, in relation to the whole data without any feature extraction applied.

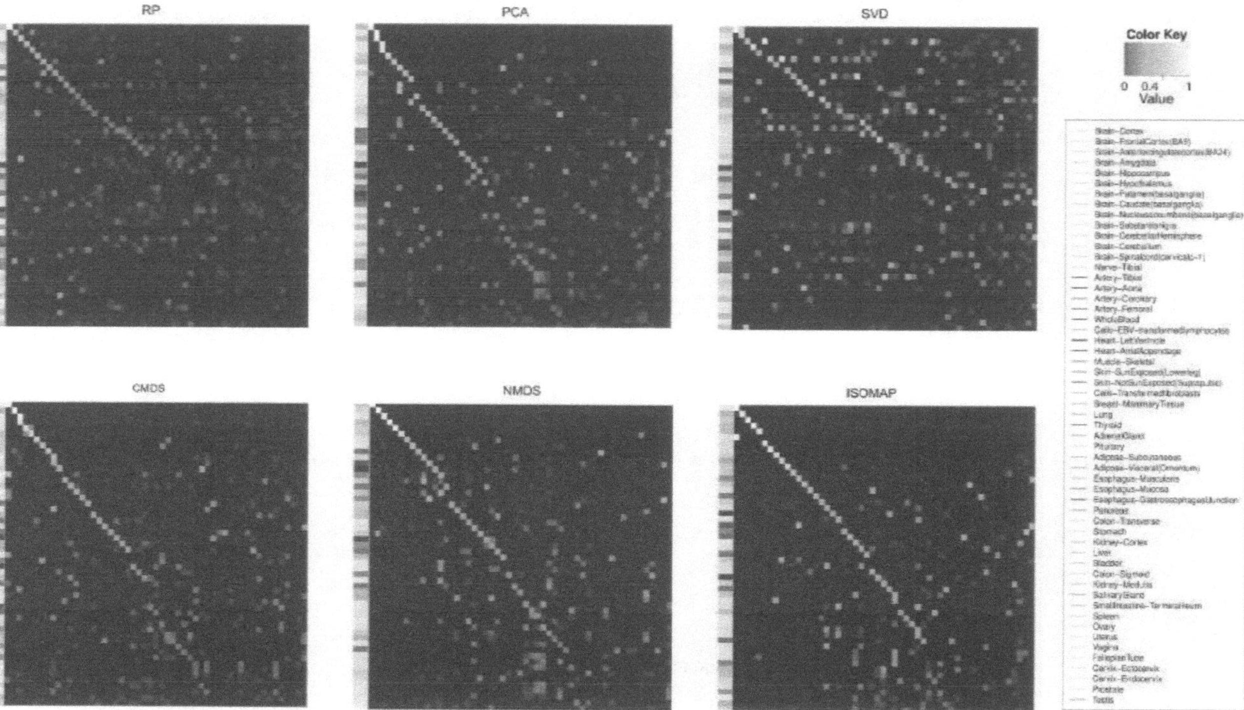
RP
PCA
SVD
CMDS
NMDS
ISOMAP
Color Key
0   0.4   1
Value

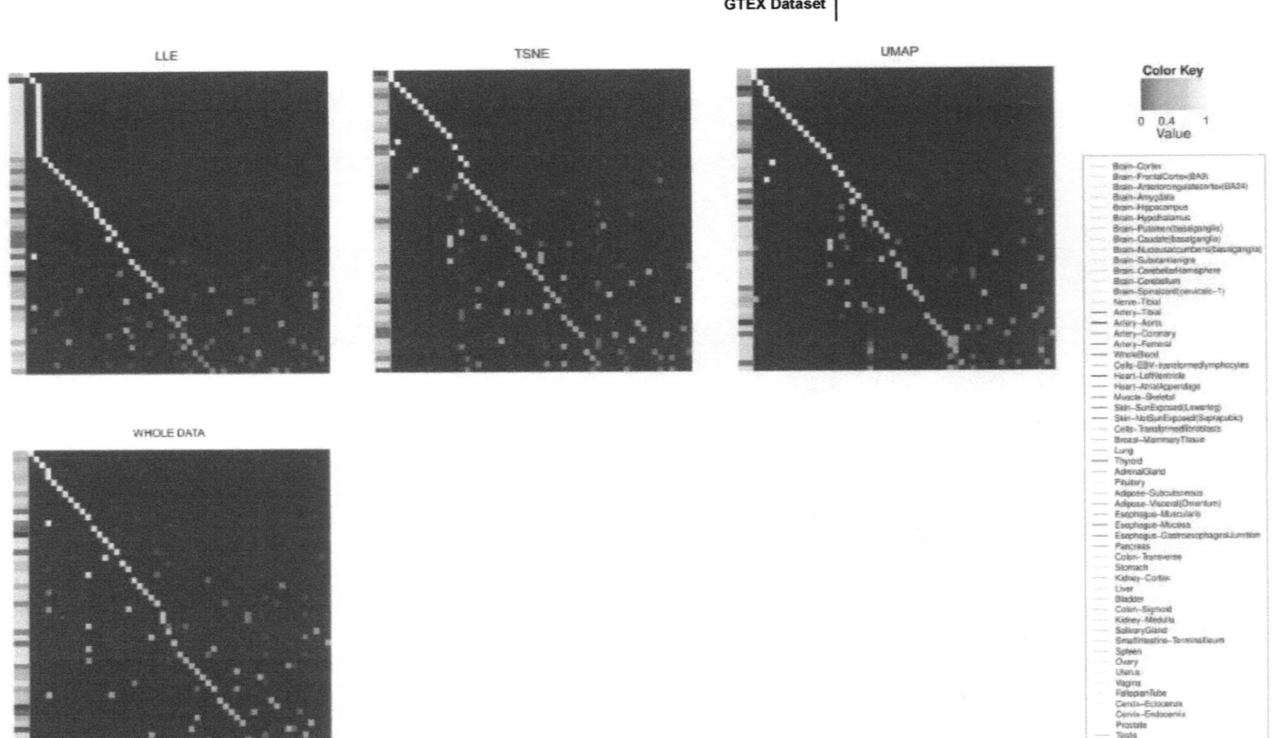

Figure 12 – Distribution of percentage of samples per tissue by the assigned clusters in two-dimensional data based on the k-means clustering. Tissues are ranked in descendent order from top to bottom by the frequency of correct assignments to their top cluster. A noticeable improvement can be visualized from the linear methods to the non-linear methods, especially for t-SNE and UMAP.

5.2.3 Summary

(a) Generally, non-linear methods demonstrate a considerably superior performance in both types of evaluation.

(b) There is not a unique method that performs the best for all the metrics evaluated. However, UMAP, t-SNE and LLE generally show the best performances for the qualitative and quantitative evaluations, as samples are more explicitly grouped. Particularly, t-SNE and LLE stand out.

(c) Improved outcomes are still notoriously obtained up to only three dimensions with t-SNE and UMAP in relation to the full dataset use, and up to four dimensions with LLE.

(d) UMAP was the method that seemingly presented the best visualizations outcomes, presenting more compact clusters and the best separation by tissue type. This result was corroborated by the quantitative evaluation.

(e) CMDS and NMDS were the non-linear methods which showed considerably inferior performances, and RP was the method which showed worse results overall.

(f) PCA was the linear method which most consistently showed better results. Additionally, it even demonstrated a better performance than CMDS and NMDS in the quantitative evaluation, mainly when up to three dimensions were used. However, in the qualitative evaluation, CMDS and NMDS showed more cluster compactness.

Chapter 6

Single-cell Dataset

6.1 Single-cell data particularities

In single-cell data analysis, the identification of cell types based on transcriptome is appealing as it allows a data-driven, coherent, and unbiased approach. DR and clustering are frequently used in the exploratory data analysis stage to identify specific patterns, such as heterogeneity across cells. However, it is very important to take special consideration for possible confounding factors such as technical noise, and the possibility of transient biological states that can disguise the underlying cell identity. The dimensionality of a human single-cell RNA-Seq data can reach up to 25,000 genes. But, in addition to the non-informativeness of several of these genes, and contrarily to the RNA-Seq data from bulk cell populations, single-cell RNA-Seq data comprise high levels of dropouts (several genes mostly contain zero counts) and noise. The dropouts problem may be due to an actual absence of transcript, a low-sequencing depth, or a failure in the library to amplify or capture the transcript. This can lead to a great deviation in normal distribution which can affect some methods' performance. Therefore, these zero counts should be filtered. However, even then, the dimensionality can be over 15,000 genes. Besides, an identified cluster may not have a biological meaning in this data, but instead be of technical origin. Thus, each cluster must be adequately verified and analysed. As there are no fixed guidelines for this verification, it may often be a slow process which requires an adequate search in literature and different databases. The use of an arrangement with known cell types by other ways such as tissues that are very well studied, or distinct cell lines, is the current best strategy available to evaluate a clustering method for this type of data [12].

6.2 Data source and pre-processing

For the single-cell data analysis, a publicly available RNA-Seq data of cardiac human embryo cells in various stages of development was used. This data is available in the National Center for Biotechnology Information Gene Expression Omnibus (GEO) with the accession ID GSE106118. There have been studies that used scRNA-Seq data to study the spatial and temporal systems of the mouse model heart development, which enabled a better understanding of the gene expression network for the organ development [87][88]. Yet, the differences between the human and mouse heart development are vast. Insights from human heart development may not only clarify the mechanisms involved in this process, but also disclose important information to better understand congenital heart diseases and cell regeneration. As samples are scarce for this data, a complete gene expression and regulatory dynamics of the human heart development is still to be clarified. Therefore, the use of DR methods for this data is, evidently, of special importance to counteract the curse of dimensionality.

This dataset had already been normalized in TPM, logarithm transformed and scaled by a colleague, but in this case using the Seurat R package [89]. Specifically, 4902 cells were sourced from 18 human embryos, ranging from 5 to 25 weeks of gestation. The total number of represented genes is 23291. After empirical evaluation of the visualizations of low-dimensional subspaces and of k-means-based evaluation metrics of low-quality cell filtered data and non-filtered data, non-filtered data was selected, since it showed improved and more consistent results. The dataset comprises 10 types of cells, namely b cells, cardiomyocytes, endothelial cells, erythrocytes, fibroblasts, granulocytes, macrophages, monocytes, NK cells and T cells. The number of instances for each class is uneven. It ranges between 1 as the minimum number of cells by cell type (erythrocytes) to 2489 as the maximum number of cells for cell type (cardiomyocytes), $\cong 17$ being the approximate median value. The number of cells per cell type is represented in Supplementary Figure 14. The cumulative variance screeplot of variance explained for this data is represented in Supplementary Figure 26. In this plot, it is illustrated that 2409 PCs are required to explain 85% of the data variance for this dataset, which is clearly a much higher number of required PCs than for the previously analysed GTEx dataset. However, the regular scree plot of variance explained shows an 'elbow' between PCs 5 and 11 (Supplementary Figure 25). The first 5 PCs were considered. The cell type annotation was used to generate the labels of the cell type for each correspondent cell.

6.3 Results

In this section, the qualitative results on the various visualizations generated for scRNA-Seq dataset is presented, as well as the respective quantitative results of the k-means clustering-based evaluation.

6.3.1 Qualitative evaluation results

The visualization results of the low-dimensional subspaces generated for the scRNA-Seq dataset of embryo heart cells are represented in Figure 13. For this data, we can also generally observe segregation by cell type for the various DR methods in analysis, particularly for cardiomyocytes, fibroblasts, endothelial cells, and macrophages. However, in general, the other cell types do not separate in a clear way. It is also evident in these graphics that nonlinear DR methods present more clear separation between the different classes of cell types in analysis.

By observing the linear methods' representations and ISOMAP and LLE methods, some outlier cells are evident. After a cross analysis of these representations with the stage of development metadata information, we verified that they are mainly fibroblasts of later development stages (20-25 weeks). Also, as expected, SVD and PCA present similar results. The visible difference is that SVD seems to aggregate clusters in a more compact way than PCA. Even though much more PCs are necessary to explain 85% of data variance in this dataset in relation to the previously analysed GTEx dataset, we can notice that, remarkably, the first two PCs comprise about 30.92% of data variance explained – 17.41% for the PC1, and 13,51% for PC2. From this method representation, we can see that a diagonal line could be drawn between cardiomyocytes and fibroblast cells, indicating that both PC1 and PC2 account for a similar part on these tissue types' separation. The third PC, represented in Supplementary Figure 15, 17 and 19 separates more visibly the macrophages tissue type, and the fourth PC represented in Supplementary Figure 16, 18 and 19, separates even more the macrophages. This component also notably separates the other cell types which are mostly not well-visible for other visualizations. Furthermore, RP visualization shows, once again, very poor results in comparison with the other representations, demonstrating much difficulty in separating the different classes of cell types present in this dataset, even though some patterns can be identified.

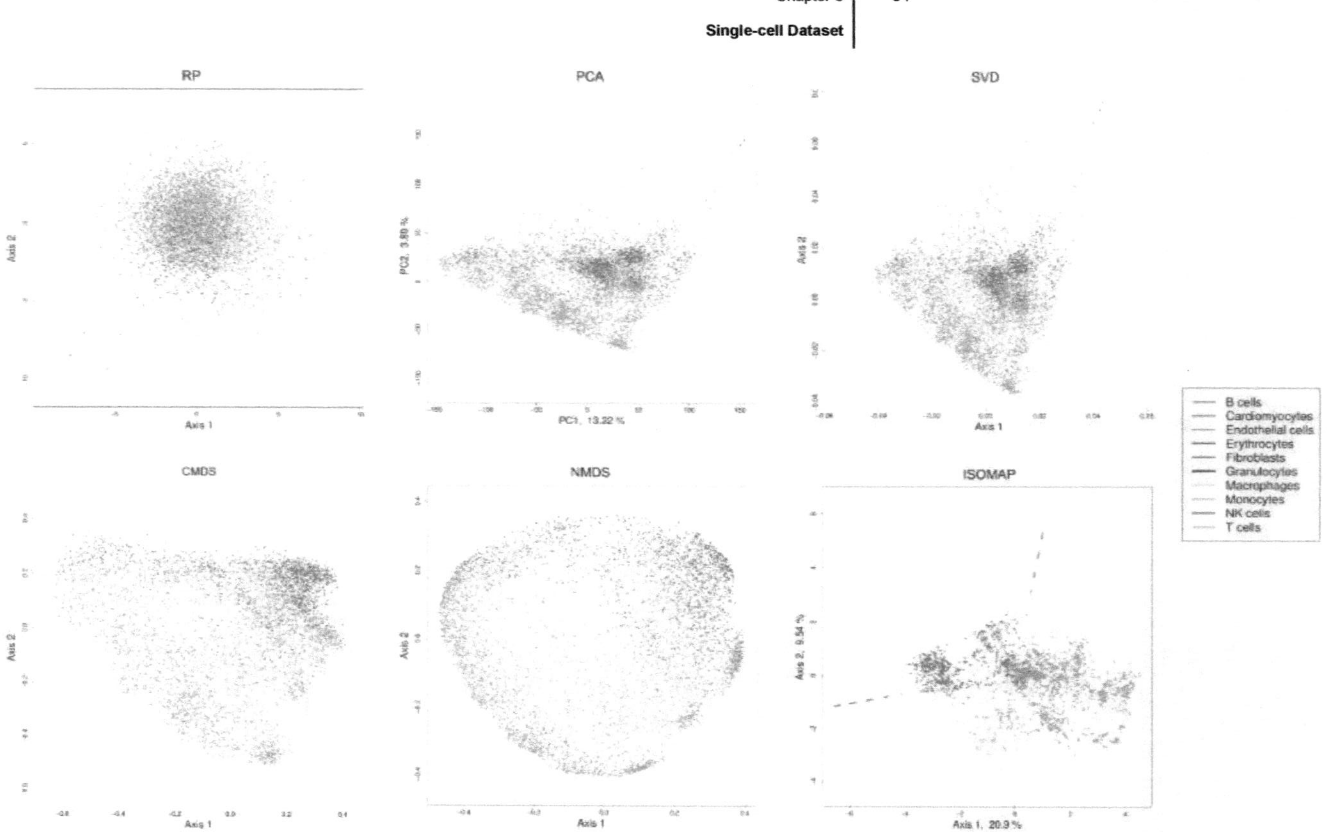
RP
PCA
SVD
CMDS
NMDS
ISOMAP
B cells
Cardiomyocytes
Endothelial cells
Erythrocytes
Fibroblasts
Granulocytes
Macrophages
Monocytes
NK cells
T cells

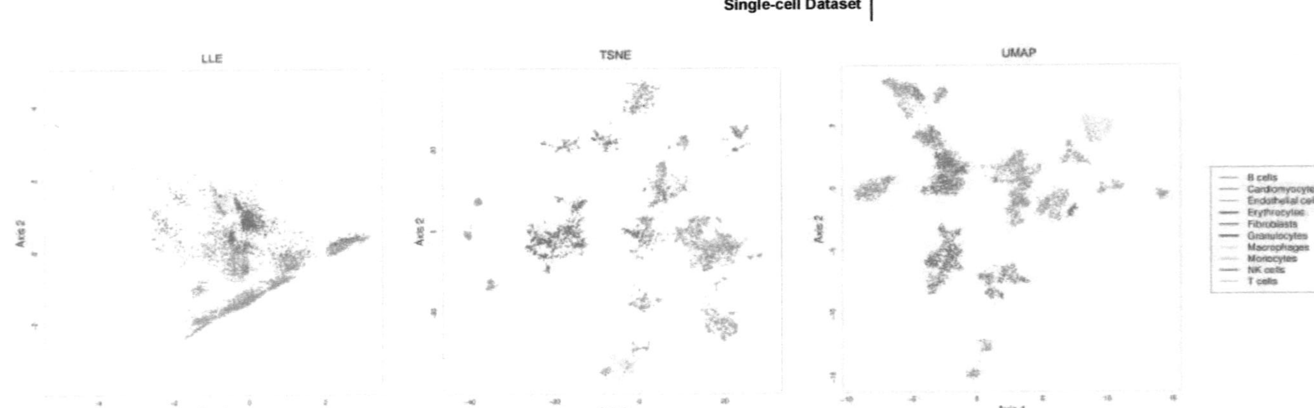

Figure 13 - Results for two dimensions for all DR methods performed on the scRNA-Seq dataset. A segregation by some cell types can be seen, especially for cardiomyocytes, fibroblasts, and macrophage cells.

For non-linear DR methods, CMDS and NMDS show much sparser results. Yet, the data points distribution of NMDS visualization seem to unravel a clearer separation of cell types for this algorithm in relation to CMDS. ISOMAP and LLE visualizations depict relatively similar representations, although more cluster tightness can be observed for LLE. For this data, Isomap presents a similar data variance explained in the first two axis in comparison with PCA. The total variance explained accounts for 30,44%, which translates in 0,47% less data variance explained in relation to PCA, to be exact. LLE representation exhibits clustering of cells in a denser way than Isomap, but t-SNE and UMAP are the methods that show an overall clearer separation of the cell types, with good cluster compactness. UMAP seems to better segregate some particular cell types, such as macrophages, even though in t-SNE the other cell types with less representativity are slightly more separated and noticeable. Moreover, cluster compactness is again more pronounced on UMAP rather than t-SNE visualizations, which shows more sparse results.

By observing the three-dimensional plots for PCA, t-SNE, and UMAP (Figure 14), the separation by cell type is, once again, even more discernible. While in PCA representation the cells are clumped, in t-SNE and UMAP representations they are notoriously separated and, inclusively, distributed in several compact subclusters. The previously observed difference between t-SNE and UMAP in the two-dimensional plot of a clearer separation of the other cell types further than cardiomyocytes, fibroblasts, and macrophages, also stands out in these three-dimensional representations.

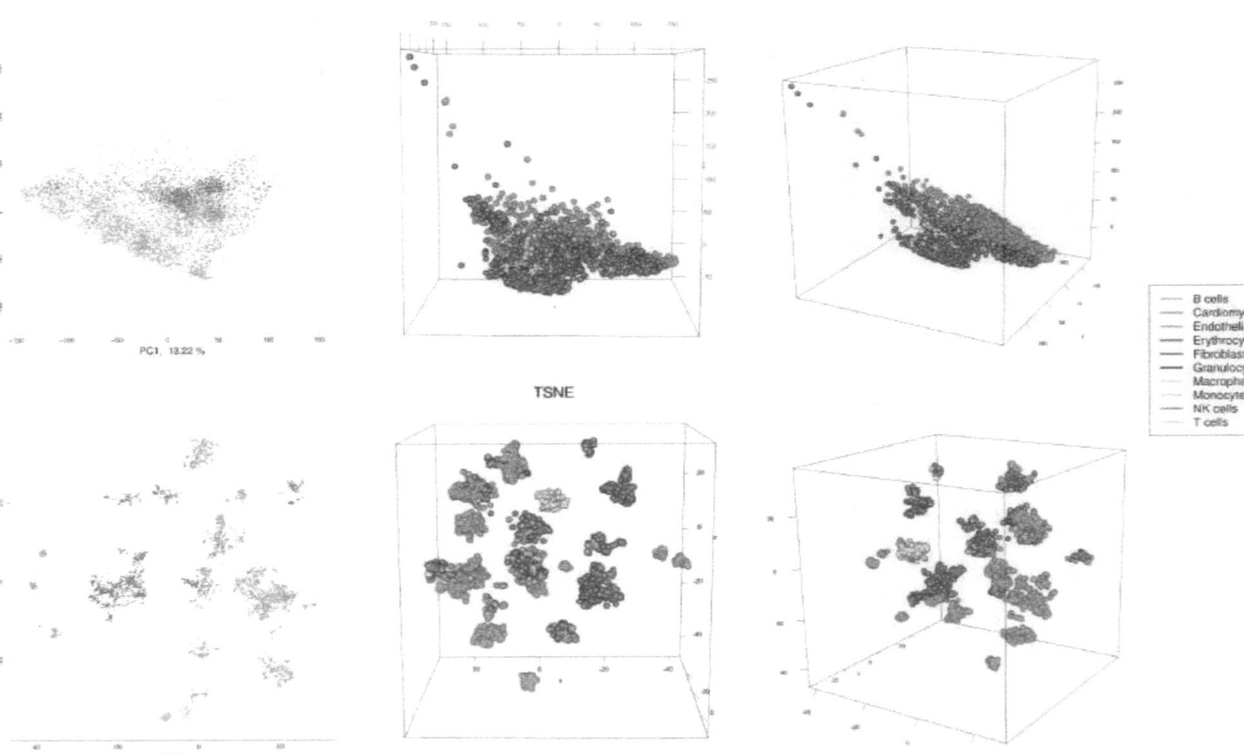
PCA
TSNE
PC2 3.69 %
PC1 13.22 %
Axis 2
Axis 1
B cells
Cardiomyocytes
Endothelial cells
Erythrocytes
Fibroblasts
Granulocytes
Macrophages
Monocytes
NK cells
T cells

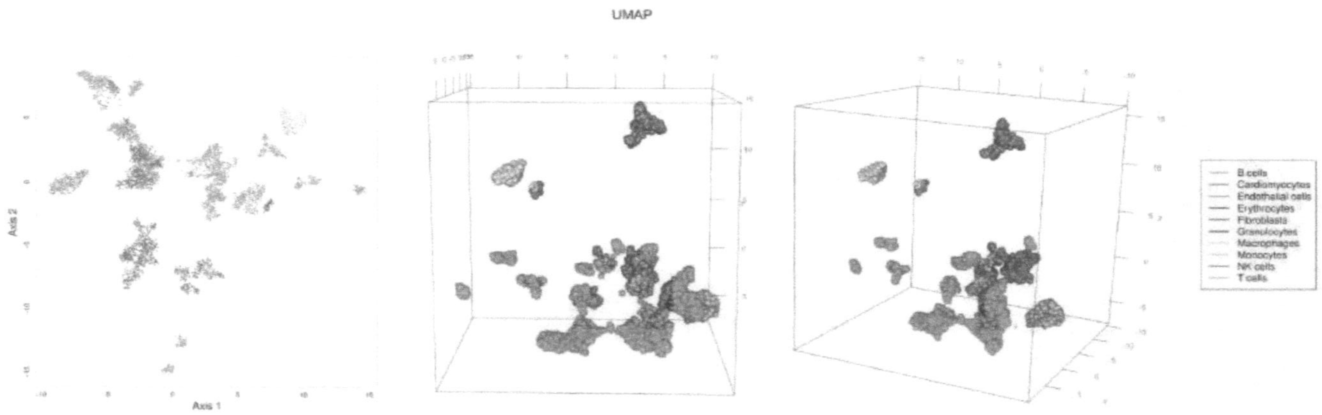

Figure 14 - Visualizations of PCA, t-SNE and UMAP representations for the scRNA-Seq data in the first two dimensions (left) and in three dimensions, on two different angles (right). A segregation by cell type is even more prominent in this 3D space. While in PCA representation the cells are clumped, in t-SNE and UMAP they are notoriously separated. Inclusively, samples seem to be distributed in several compact subclusters in these latter two methods. Samples are coloured by their cell type.

Examining the Hierarchical clustering visualization (Figure 15), a segregation by cell type is also visible. There are various separated clusters within each tissue type as well, indicating a clear subcluster pattern. These results are consistent with t-SNE and UMAP qualitative evaluations, in which it is also noticeable, generally, the existence of clear subclusters for some tissue types. As we are analysing a heavily imbalanced data, which is a typical occurrence in gene expression data and particularly in single-cell data, the cell types which have more representativity in the dataset are, of course, more easily visualized. Cardiomyocytes and fibroblasts are the cell types which clearly present various subclusters. For other cell types, it is not possible to analyse.

A question that may arise, is if the observation of these subclusters is related to the different stages of development. To analyse this hypothesis, we examined the correspondent visualization results of PCA, t-SNE and UMAP and the k-means clustering based evaluation metrics with the main developmental stages as the class labels. However, even though some arrangement was observed, we did not verify an evident pattern. We presume that these subclusters may be rather due to the intricate cardiogenesis process, which involves various synergistic actions of different cell lineages. However, righteous conclusions on this subject could only be taken with the posterior analysis on the resulting subclusters by the study of a spatial and temporal analysis.

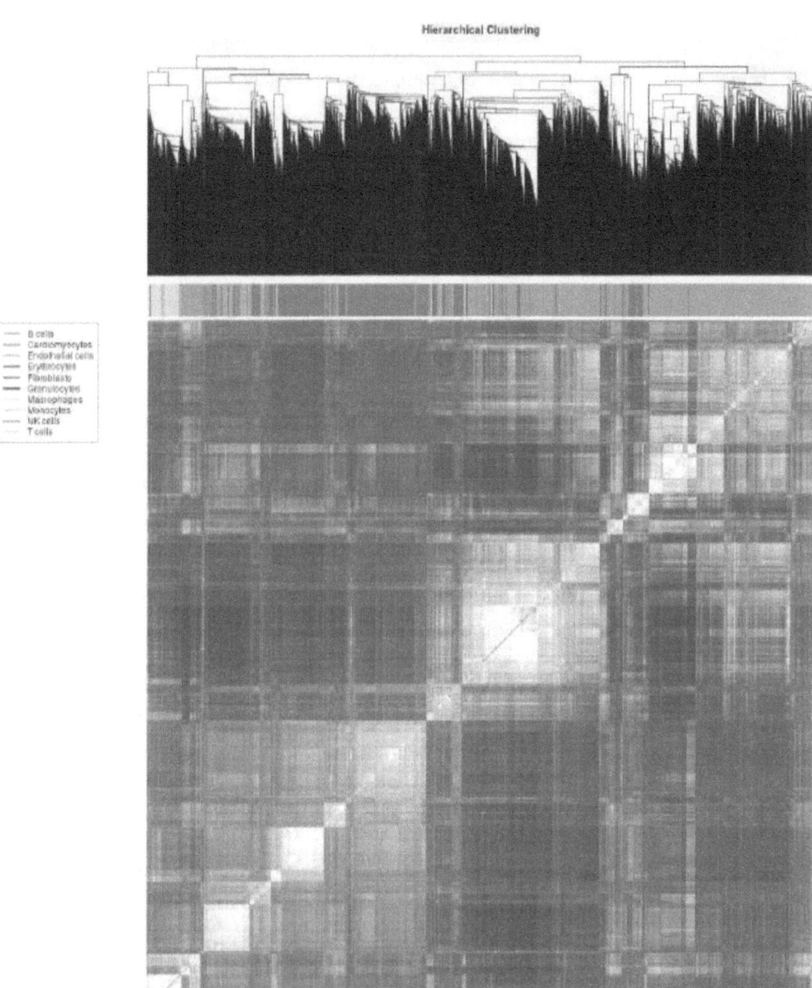

Figure 15 - Hierarchical Clustering visualization result for scRNA-Seq dataset. A segregation by tissue can clearly be visualized, as well as some subclusters within some tissues, generally corresponding to a specific part of the tissue in detail. Correlation between samples is represented by a colour code from blue (negative correlation) to red (positive correlation). White cells represent that there is no correlation between samples.

6.3.2 Quantitative evaluation results

Figure 16 shows the quantitative evaluation results for this dataset. From these, it can be immediately noticed that the differences among the clustering metrics for the various DR approaches are not so pronounced as in GTEx dataset, with no striking differences in classification and clustering validity measures for the various methods. Additionally, it is also noticeable a much inferior performance in comparison. However, there are some results that are common for both datasets. First, there is an observed improvement in performance for the non-linear methods, especially for ISOMAP, LLE, t-SNE and UMAP, in a much less prominent way. Second, is the worst performance of the RP method among all the methods applied. As expected, all methods seem to perform better with higher number of dimensions, even though the differences in performance between each considered number are less pronounced when compared to the GTEx dataset. These latter observations corroborate that, in contrast with the GTEx dataset, the first dimensions comprise much less variance (Supplementary Figure 26). These results are also in concordance with the known and previously discussed characteristics of single-cell data in general, in comparison with bulk RNA-Seq data. Importantly, LLE and UMAP projections up to two dimensions still show approximate results for the considered metrics compared to when the full dataset is used. In this case, t-SNE reveals more difficulties than these methods.

For the distribution of samples per tissue by the assigned clusters in two-dimensional data, a noticeable improvement can also be visualized from the linear methods to the non-linear methods (Figure 17). This improvement is more pronounced for t-SNE and UMAP representations, which show a similar distribution to the whole data.

Heatmaps on Supplementary Figure 20, 22 and 24 show that the ISOMAP, UMAP, and t-SNE are the methods which more closely approximate to the full dataset results. Once again, NMDS surpasses the CMDS performance, but only for a slight margin. The linear method which demonstrates most correct class assignment is the SVD method, but with not much advantage over PCA. The results in three dimensions are generally in accordance with these mentioned two-dimensional results. However, in the four-dimensional space, all the methods' performances are closely similar, and even with a slight advantage for the linear methods over the non-linear ones.

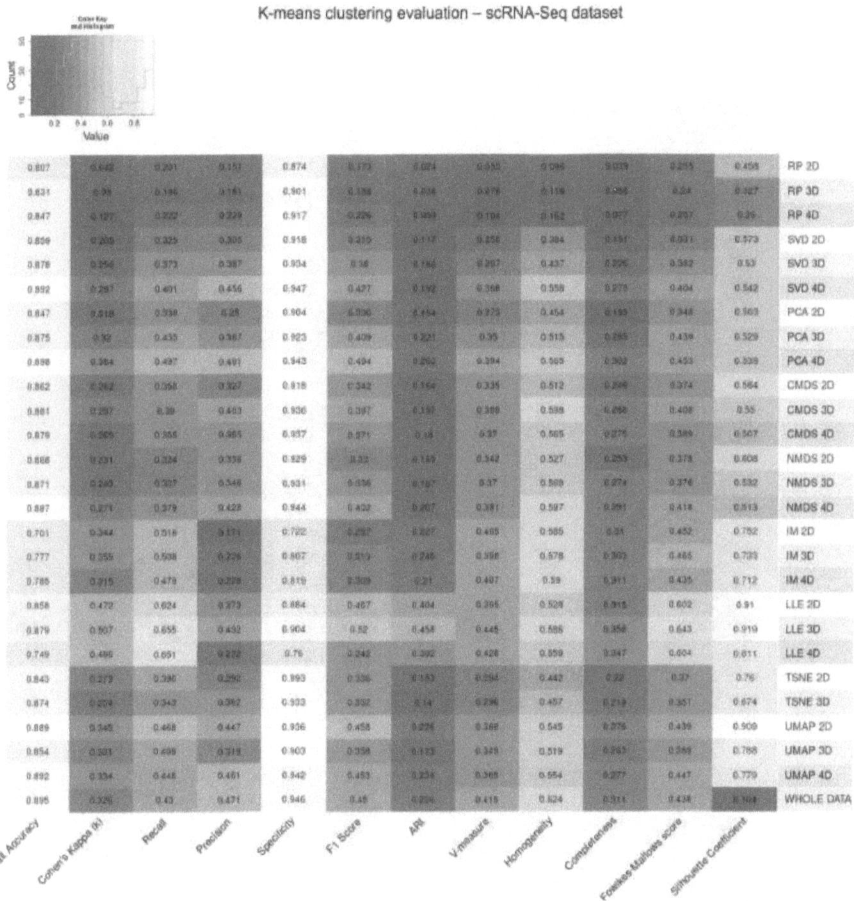

Figure 16 - K-means clustering evaluation metrics results for the scRNA-Seq dataset. Values range from 0 to 1, in a colour gradient from red to yellow, respectively. A slight improvement from the linear methods to the non-linear methods can be observed. It is also noticeable a much inferior performance in clustering validity and classification metrics results in comparison to the GTEx dataset.

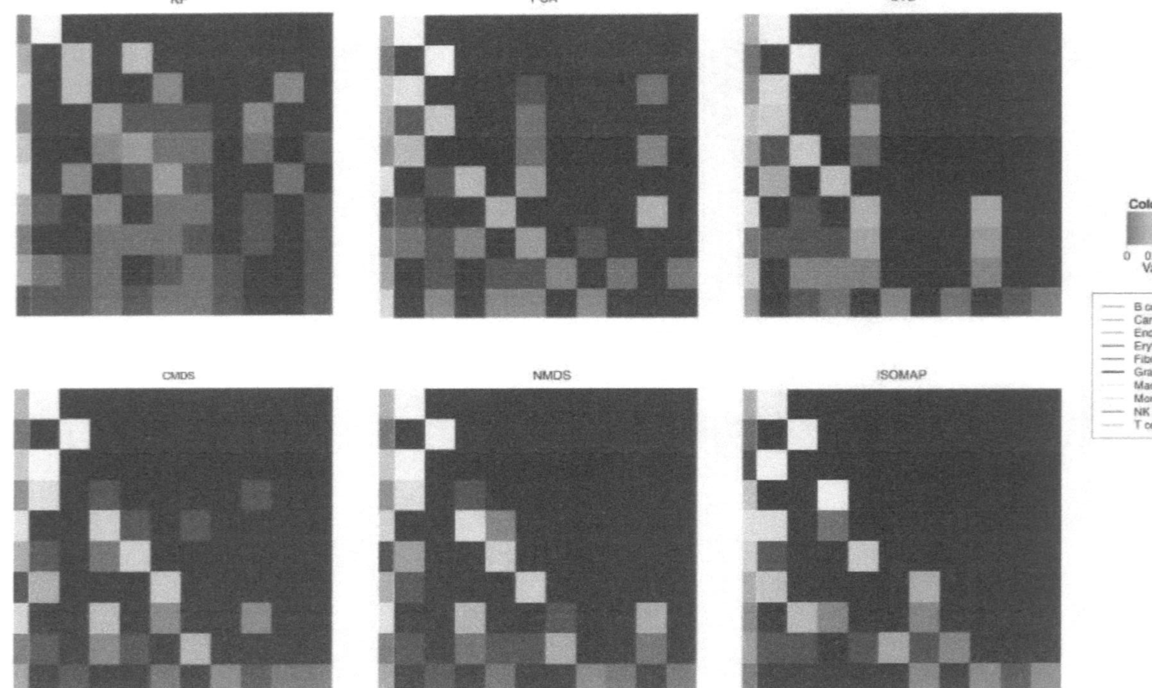
RP
PCA
SVD
CMDS
NMDS
ISOMAP
Color Key
0   0.4   1
Value
B cells
Cardiomyocytes
Endothelial cells
Erythrocytes
Fibroblasts
Granulocytes
Macrophages
Monocytes
NK cells
T cells

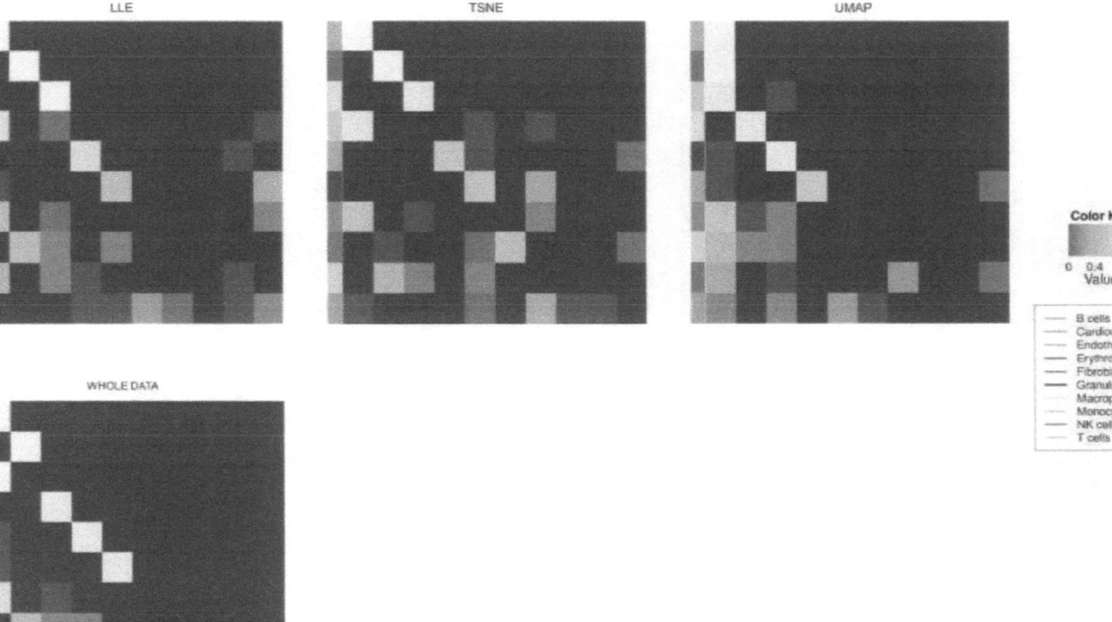

Figure 17 - Distribution of percentage of samples per tissue by the assigned clusters in two-dimensional data based on the k-means clustering. Tissues are ranked in descendent order from top to bottom by the frequency of correct assignments to their top cluster. A noticeable improvement can be visualized from the linear methods to the non-linear methods, especially for t-SNE and UMAP.

6.3.3 Summary

- In consistence with GTEx dataset results, non-linear methods also generally demonstrated better results in both types of evaluation for this dataset, although in much less difference. Among these methods, CMDS and NMDS were again the ones which presented more difficulties.

- Also, and even more markedly, these outcomes substantiate that there is not a single method that outperforms others in all the metrics considered. Yet, better general performances are seen for UMAP, t-SNE, LLE and Isomap. Once again, t-SNE and LLE show most consistent results.

- Best visualization outcomes are again seen for UMAP method in terms of compactness and segregation by cell type, although cells with much less representativity are more easily observed in t-SNE representations.

- Linear methods demonstrated mostly similar results between them, with a slight advantage for SVD. RP again indicated a worst data representation among all the methods.

- Similar quantitative outcomes are still remarkably obtained up to only two dimensions with LLE and UMAP reduced spaces in relation to the full dataset space. This time, t-SNE was not able to achieve equivalent results in this perspective.

- A subcluster pattern can be noticed in t-SNE and UMAP representations. This observation is corroborated by Hierarchical clustering method results. Remarkably, it likely indicates important differences within cells for the various considered human embryo cell types.

Chapter 7

Computational Demands

Though some DR methods might show better results in retrieving underlying data patterns, if they take a long time to execute and/or require much memory, it may be preferable to apply simpler ones with a higher number of dimensions. Here, the CPU time results are shown in Figure 18 (a) and (b) and the maximum memory usage in (c) and (d).

Remarkably, we can observe that t-SNE and UMAP running time was smaller than simpler methods such as PCA and SVD. This difference is even higher when comparing the former to Isomap, LLE and NMDS methods. Curiously, t-SNE executed in less time than UMAP in both datasets. Another difference that stands out is that, when compared to the GTEx dataset, LLE performed in a larger amount of time in the scRNA-seq dataset in proportion to other methods. Additionally, the RP method demonstrates the implementation which performed in a smaller amount of time, as expected, as it only requires a simple matrix multiplication. K-means clustering CPU time, in the other hand, is very similar among DR methods projections, even though for CMDS, NMDS and t-SNE in the scRNA-Seq dataset it did marginally present a larger runtime. As shown, generally, its execution on the reduced spaces requires much less time than on full datasets. In comparison, it only required a few seconds to run on the former, while on the latter it took a few days.

From the maximum memory usage graphics, it can be noticed that the latter k-means clustering time results are in accordance with its maximum memory usage. Furthermore, among the various DR methods, RP, CMDS and UMAP generally show less memory usage. Particularly, it can be observed that the ones which allocated more memory at once for the GTEx dataset were PCA, SVD and LLE, while for the scRNA-seq dataset these were LLE, NMDS and ISOMAP.

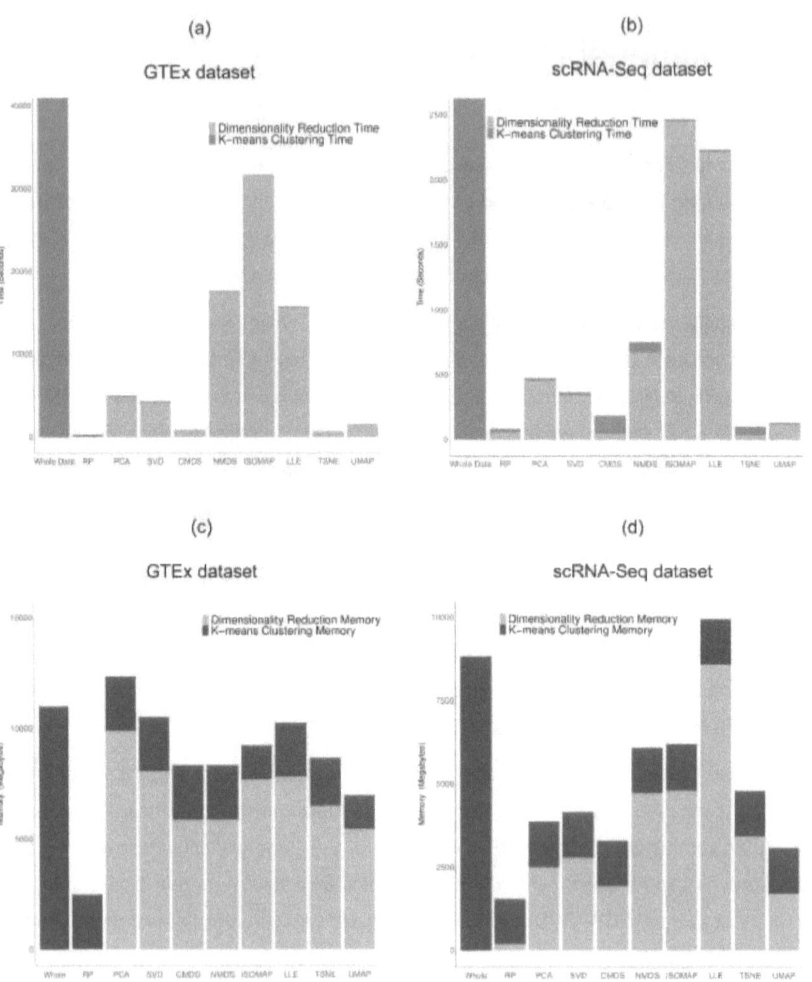

Figure 18 - CPU Time ((a) and (b)), and maximum memory usage ((c) and (d)) of each DR method and the respective k-means clustering execution. As clustering runtimes for the full datasets are much larger, they are truncated. The runtime for the whole data for the GTEx dataset was 10 days, and for the scRNA-Seq dataset it was 34 hours.

Chapter 8

Conclusions

8.1   Summary on Guideline Recommendations

(a) Prior information on the biological relevance of the various features for the structural and prediction analysis are of particular importance for feature filtering and examination of DR results. Thus, it is of critical importance to use this information whenever it is possible.

(b) Some DR methods such as NMDS, t-SNE and UMAP can benefit from a prior initialization with another simpler DR technique such as CMDS for NMDS, and PCA for t-SNE and UMAP. The selection of fewer and uncorrelated variables helps in computational efficiency, reduces noise, and helps to yield better results.

(c) Non-linear DR methods generally require several hyperparameters to tune and may return very different results for each configuration. Hyper-parameter tuning can be a tedious and time-consuming task. Rules of thumb in literature are scarce, with little agreement, and may not give the best results. Nevertheless, these empiric rules may aid in an initial phase of the analysis. Therefore, it is very important to experiment the impact of the different hyper-parameters.

(d) t-SNE is frequently the choice for the visualization of scRNA-seq data and one of the most used for the visualization of bulk RNA-Seq data. However, it does not allow an adequate interpretation of the visualization outcomes, as the distances between clusters may not have any biological significance and may neglect possible relations between cells or samples, as it focuses on preserving local structure. UMAP method, on the other hand, can preserve both local and global structure. This important characteristic allows for a suitable interpretation of the visualized patterns. For this reason, UMAP is a good alternative for this task. Additionally, UMAP is also able to reduce data in more than three dimensions, contrarily to t-SNE, and, as shown, may provide a much better representation of data in relation to other widely used methods.

(e) If metadata information is available it should also be considered for the analysis, as it may explain some of the found patterns. This qualitative verification should as well be combined with quantitative analysis for a more precise evaluation.

(f) Preferably, in practice, the reduced representation of each DR method should present a dimensionality that matches the intrinsic dimensionality of the data. This intrinsic dimensionality is the minimum number of dimensions required to explain the observed traits and structure of the data. As the main aim of this thesis was to evaluate and compare diverse DR methods in the same conditions, this approach was not adopted.

(g) We verified that the silhouette coefficient does not always positively correlate with the best possible representations of the various feature extraction methods. Thus, it is very important to consider a broad set of evaluation metrics subsequently to each method implementation in order to properly tune its hyperparameters, and thus obtain the best and most appropriate representation. Particularly, this recommendation has special relevance for comparison studies of the different DR approaches.

(h) Different configurations should be adopted for visualization and clustering purposes if necessary, as the most appropriate configuration for visualization is frequently not the best possible configuration for clustering analysis. This is especially the case for non-linear DR methods such as Isomap, LLE, t-SNE and UMAP.

(i) RP method is the fastest in execution. Adding to its very simple implementation, it may also constitute an attractive choice in an initial phase of data analysis for very large datasets. However, as demonstrated, a much less accurate representation is retrieved.

8.2 Main Contribution

Despite the enormous potential and interest that arose with the introduction of new generation sequencing and particularly with RNA-Seq, there are few comparative studies on the various DR and clustering methods that can enlighten the data exploration process and facilitate further analysis for this type of data. The application of these methods is of crucial importance for the analysis of RNA-Seq data. Yet, as some of these methods require solid hyper-parameter tuning coupled with a reasonable understanding of the theoretical aspects, there may be a tendency for researchers to rely on the most basic and common linear approaches. However, the detection of more complex patterns will require non-linear methods.

Taking these notions into consideration, we proposed an assessment that provides a comprehensive evaluation and comparison between several of the most interesting methods for genomics data analysis in order to maximize the knowledge that can be extracted from data. Our thesis may help researchers in choosing, applying, and evaluating the most appropriate DR methods for the analysis of RNA-Seq data. This consideration is based on the global outcomes, including the review on basic concepts of DR and clustering methods, the framework for a comprehensive evaluation and comparison of the different methods, including a qualitative and quantitative evaluation, and the resultant summarized guidelines. Ultimately, these outcomes may not only be useful for the analysis of bulk RNA-Seq and single-cell RNA-Seq data, but also for other types of high-dimensional data.

8.3 Future Work

In this thesis, we have primarily focused on two clustering methods, one in combination with various types of feature extraction algorithms for a clustering-based evaluation - k-means - and other in the perspective of a global examination of the data structure – hierarchical clustering. A future evaluation process with more clustering approaches could bring a more comprehensive insight on the evaluation of the studied methods.

Moreover, a straightforward gene filtering process was used. Eventually, it can be beneficial in future work to use other feature selection methods or a stricter filtering-based approach and examine its results, as it could potentially reduce the computational time for implementing the various studied methods.

A higher number of dimensions analysed may also bring us other perspectives. Yet, a similar amount of visualizations results may not be feasible or desirable. As some of the DR methods analysed, especially the manifold learning non-linear methods, are very sensitive to fine tuning hyperparameters, automatic processes of hyperparameter selection that properly consider all the drawbacks for each method can be of crucial value.

Furthermore, some methods, such as Multidimensional scaling, LLE and Isomap still present a high computational time. Therefore, more robust, and automatic procedures are needed to overcome the limitations on the implementation of these algorithms. Ultimately, they could allow a much easier and faster implementation for an overall robust analysis.

Appendix A

Supplementary Visualizations of GTEx dataset

This appendix contains all the supplementary visualizations generated for the analysis of the GTEx dataset. It comprises the following: a bar plot figure illustrating the number of samples per tissue type; Visualization results for all the DR methods of the first and third, first and fourth, second and third, second and fourth, and third and fourth dimensions; Heatmaps of k-means clustering-based evaluation recall metric in two, three and four-dimensional reduced data; Heatmaps of the frequency distribution of the samples for each tissue by the assigned clusters for three and four-dimensional reduced data; A scree plot of the variance explained for the first 100 principal components of PCA (PCs); and a scree plot of the cumulative variance explained for the first 100 principal components of PCA (PCs).

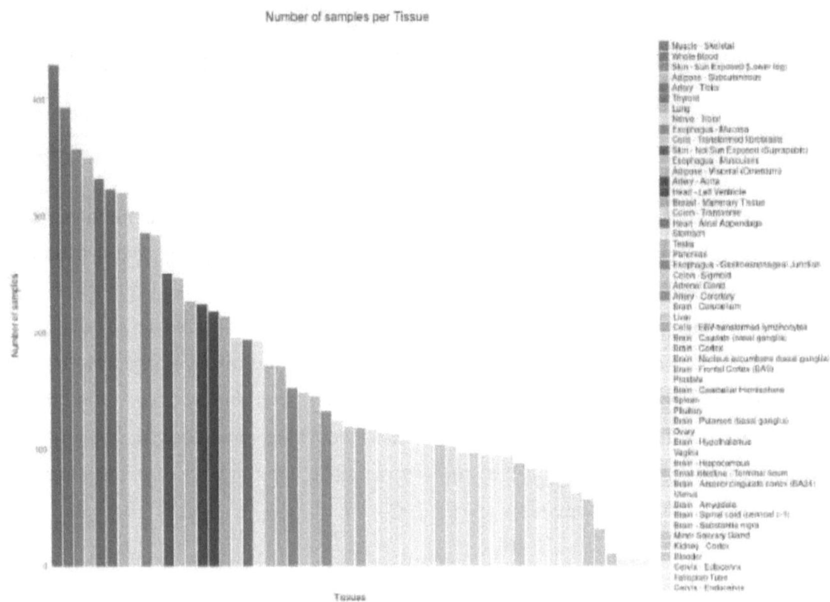

Supplementary Figure 1 - Representation of the number of samples for each tissue type of the GTEx dataset on descendent order. It ranges between 430 (muscle-skeletal tissue) to 5 (cervix - endocervix tissue). The calculated median number of samples per tissue type is 119.

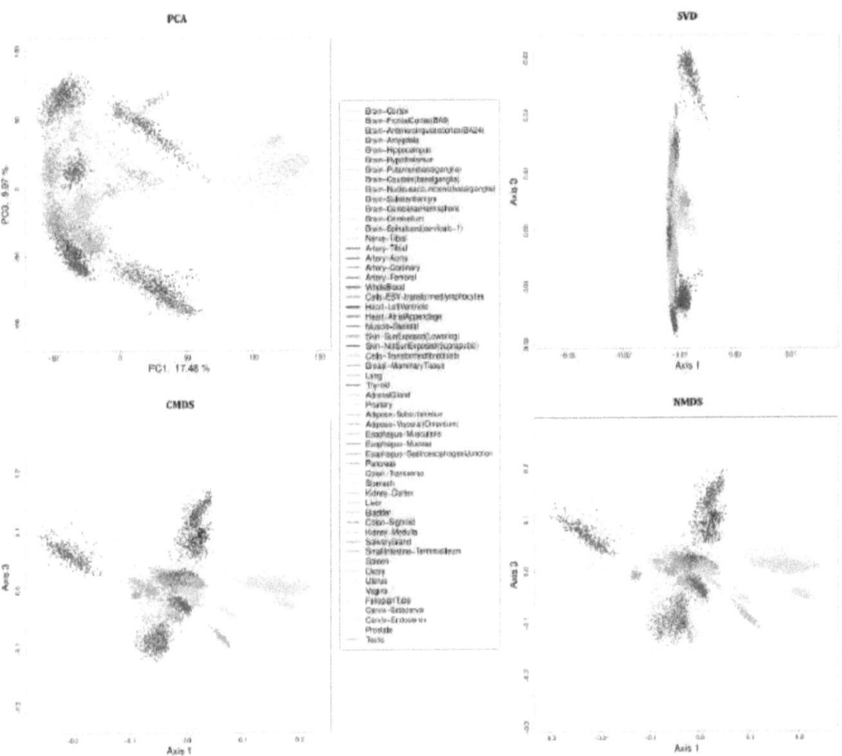

PCA
SVD
CMDS
NMDS

Supplementary Figure 2 – Visualization results of the first and third dimensions on the GTEx dataset

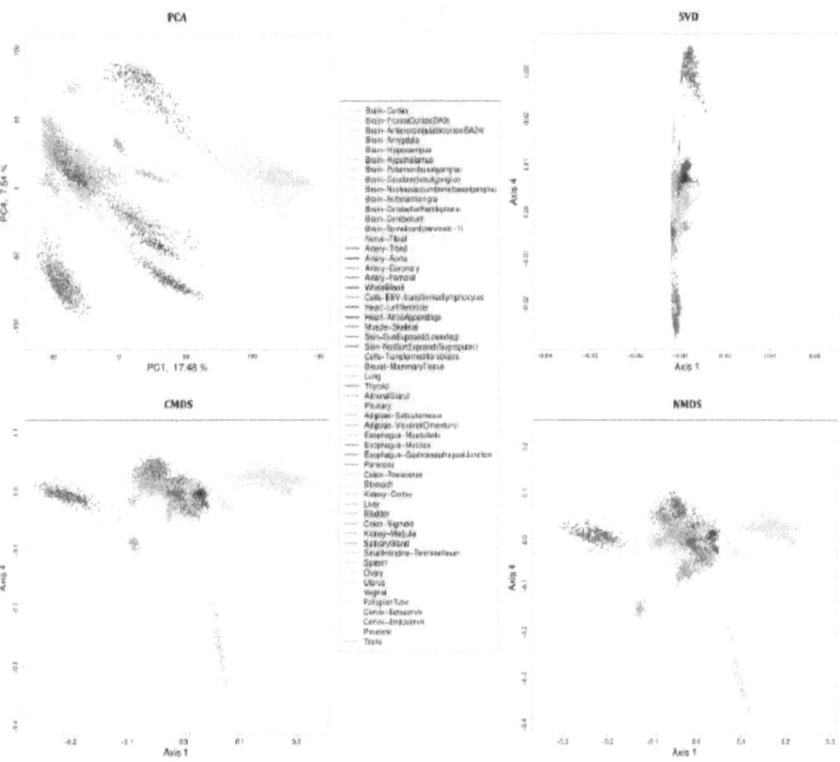

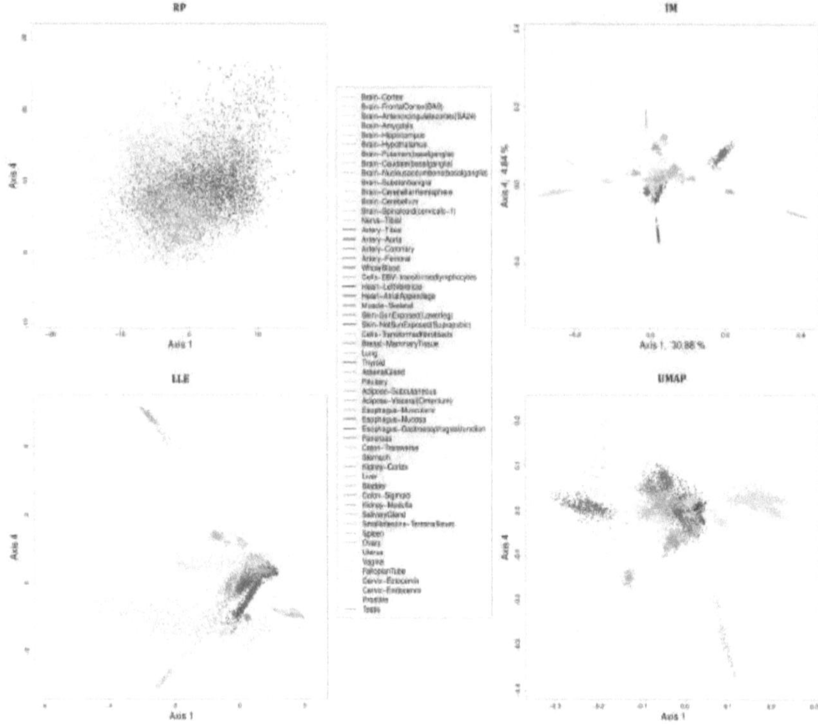

Supplementary Figure 3 – Visualization results of the first and fourth dimensions on the GTEx dataset.

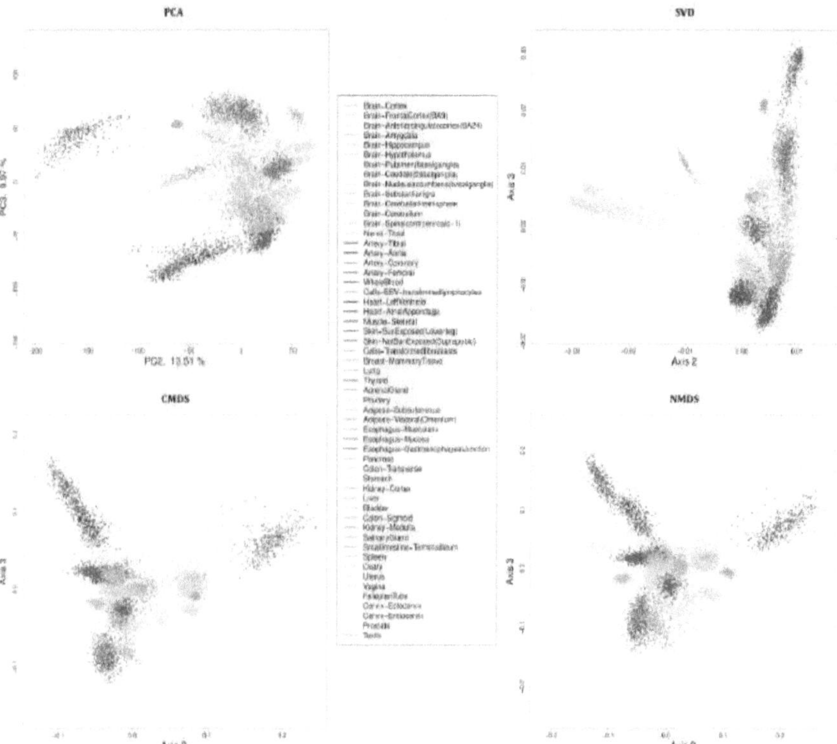

Supplementary Figure 4 – Visualization results of the second and third dimensions on the GTEx dataset.

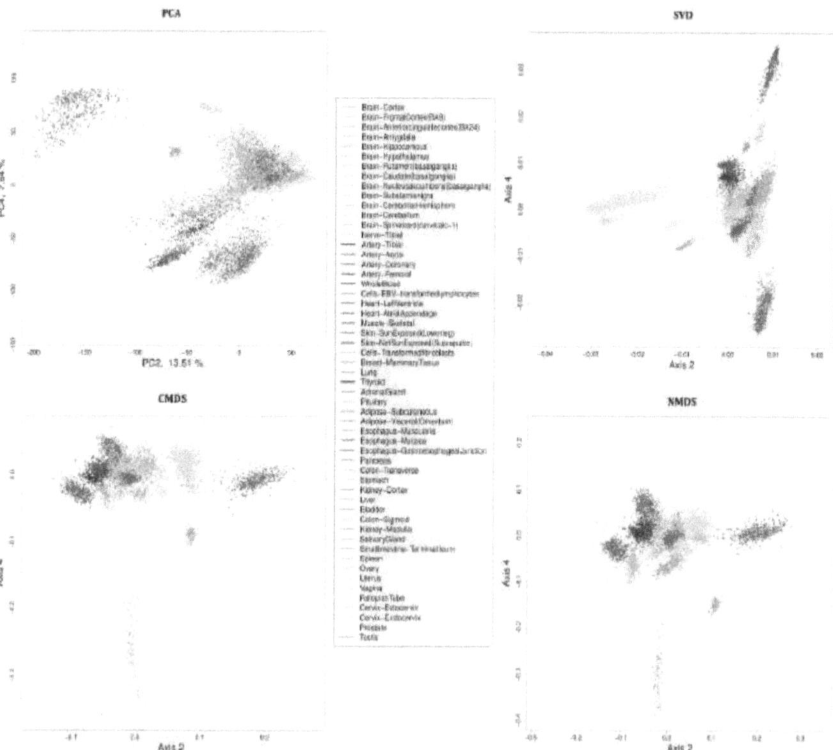

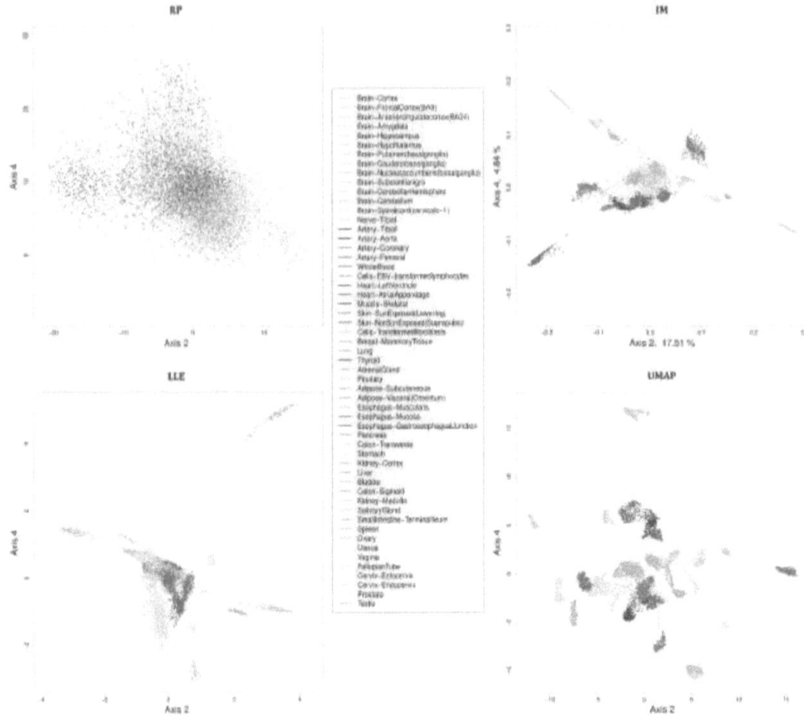

Supplementary Figure 5 – Visualization results of the second and fourth dimensions on the GTEx dataset.

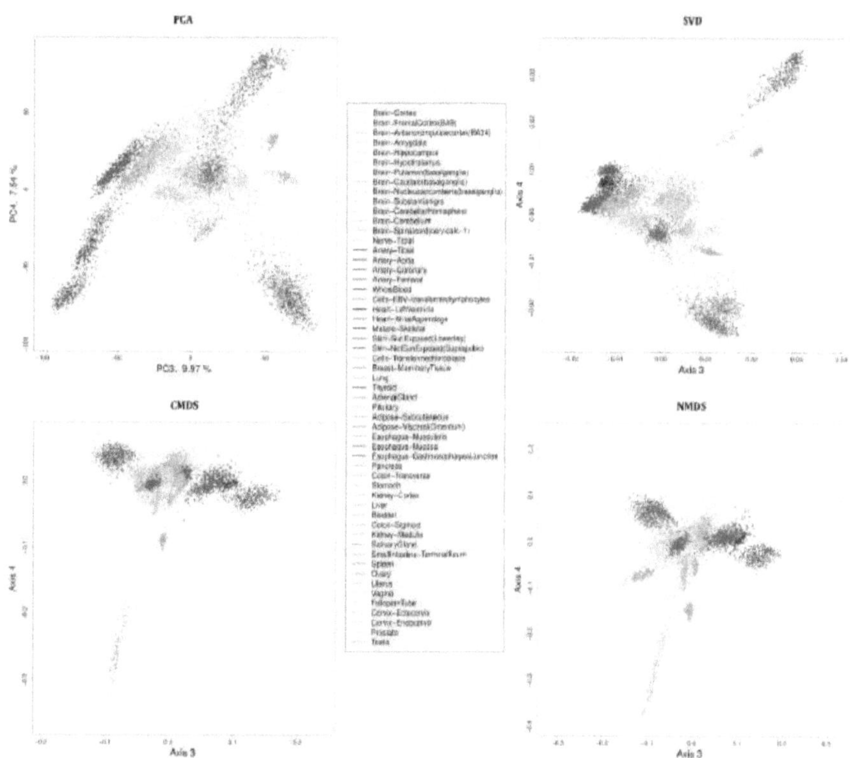

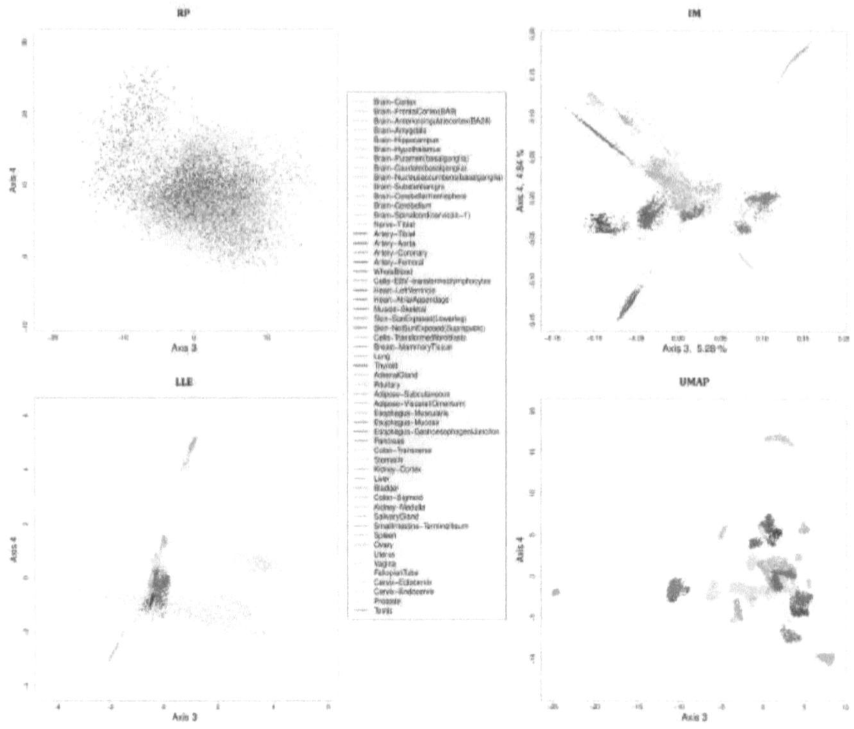

Supplementary Figure 6 – Visualization results of the third and fourth dimensions on the GTEx dataset.

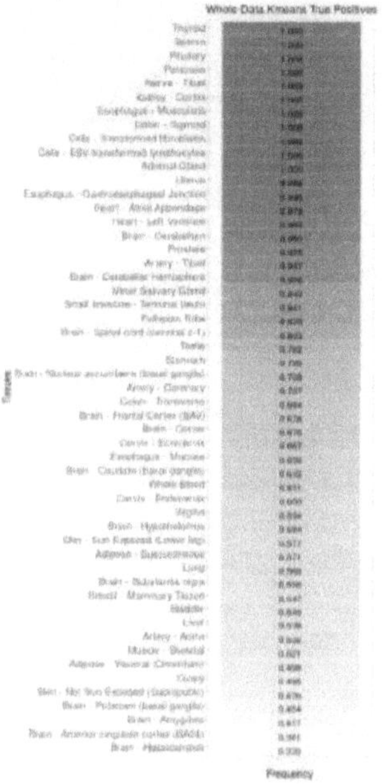
Whole Data Kmeans True Positives
Tissues
Frequency

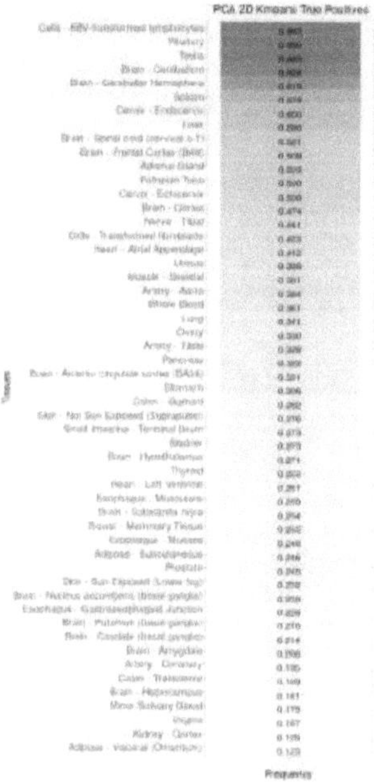
PCA 2D Kmeans True Positives
Tissues
Frequency

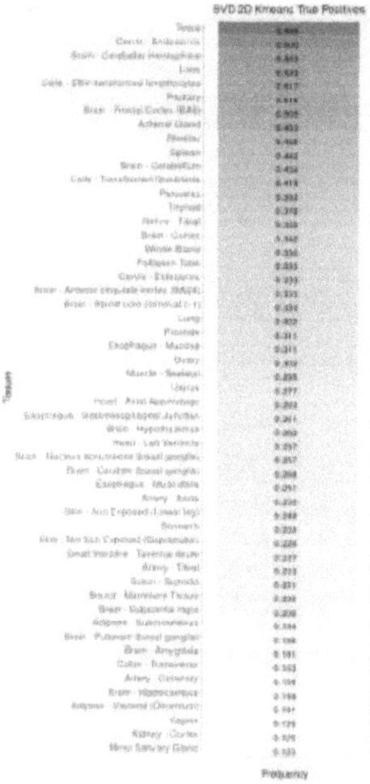
SVD 2D Kmeans True Positives
Tissues
Frequency

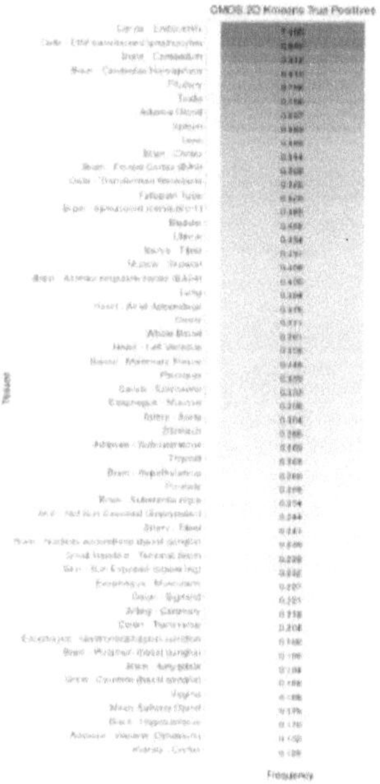
CMDS 2D Kmeans True Positives
Tissues
Frequency

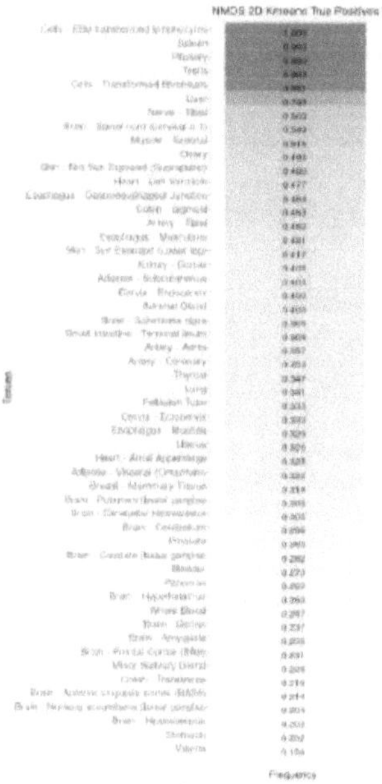
NMDS 2D Kmeans True Positives
Tissues
Frequency

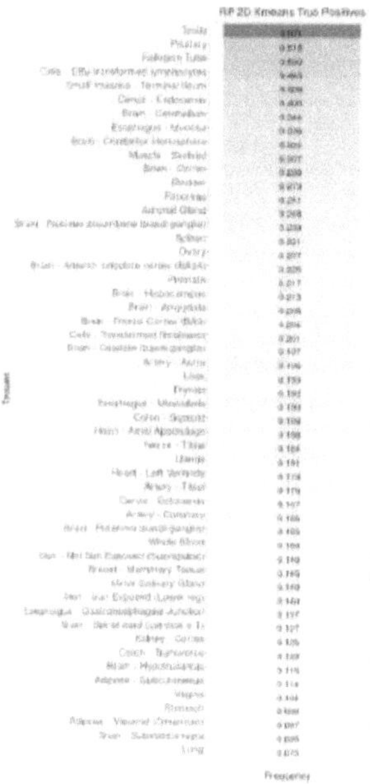
RP 2D Kmeans True Positives
Tissues
Frequency

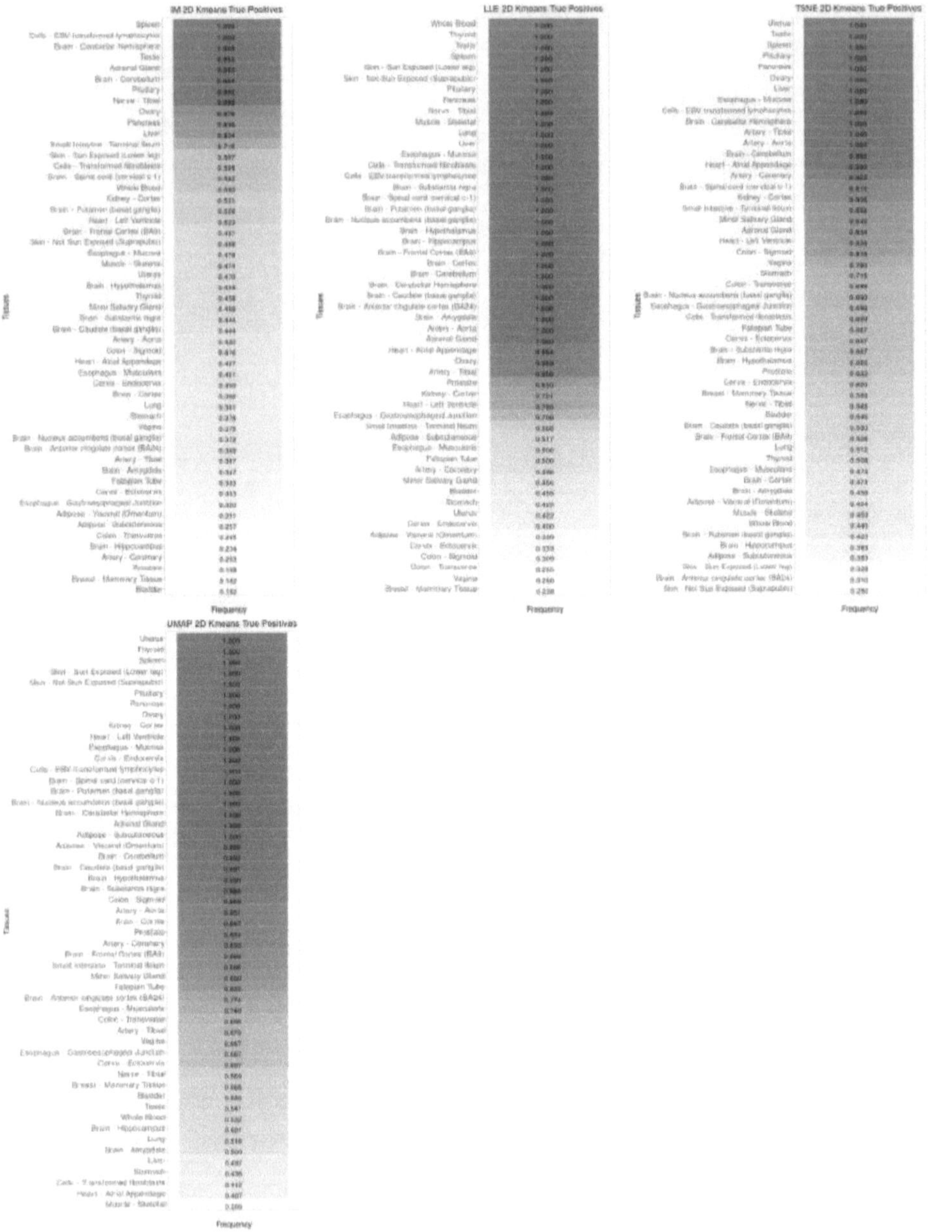

Supplementary Figure 7 – Recall rate based on the clustering evaluation for DR on two dimensions and for the full dataset.

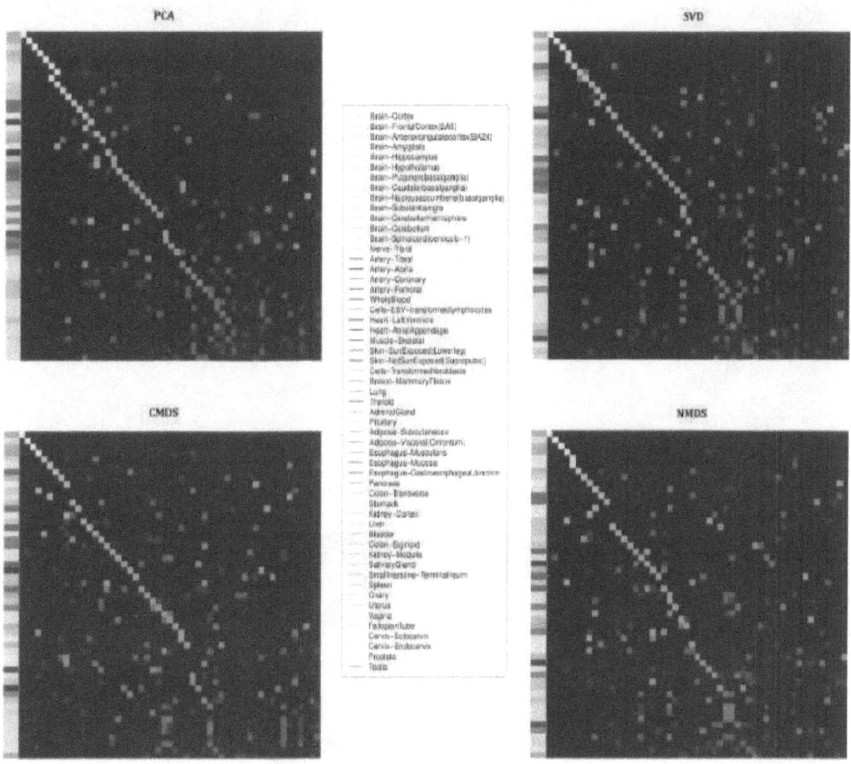

PCA
SVD
CMDS
NMDS

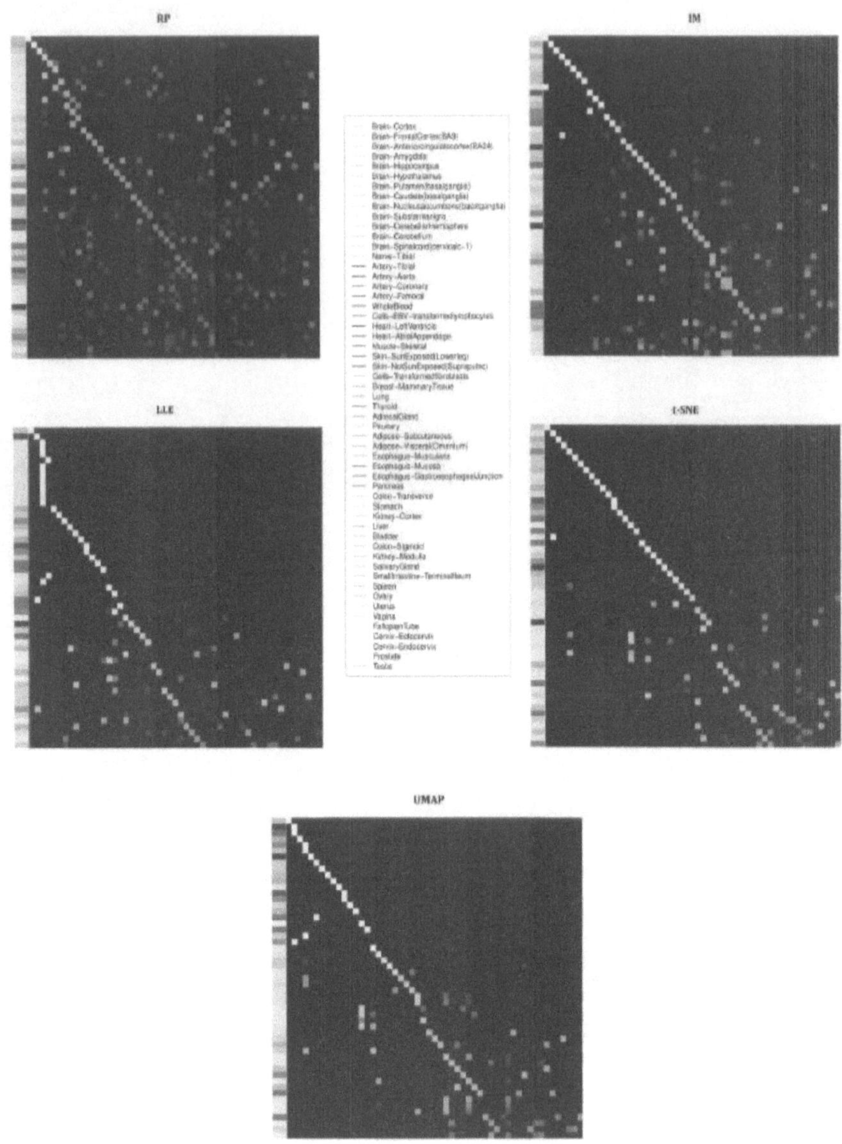

Supplementary Figure 8 - Distribution of percentage of samples per tissue by the assigned clusters in three-dimensional data based on the k-means clustering. Tissues are ranked in descendent order from top to bottom by the frequency of correct assignments to their top cluster.

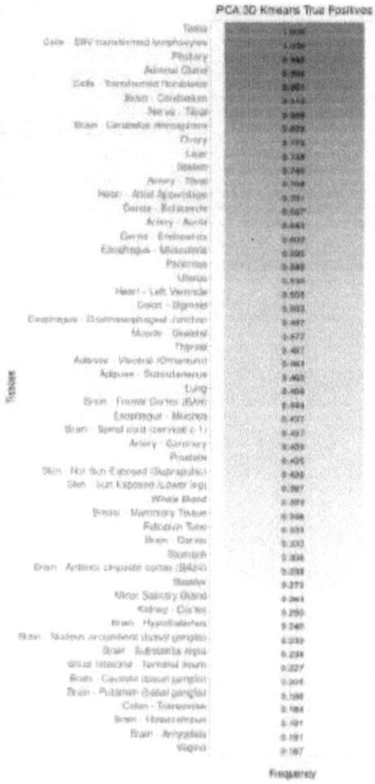
PCA 3D Kmeans True Positives
Tissues
Frequency

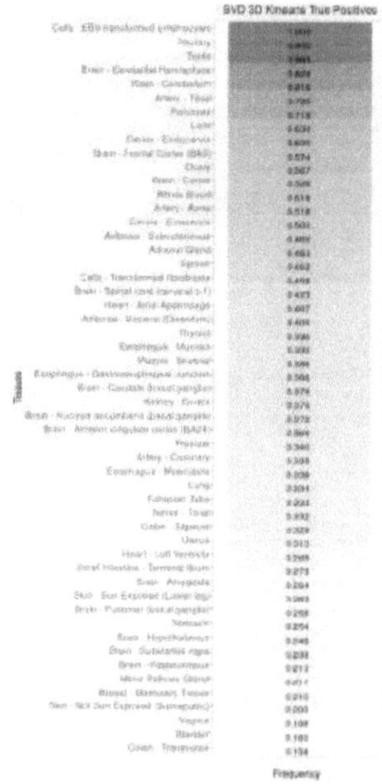
SVD 3D Kmeans True Positives
Tissues
Frequency

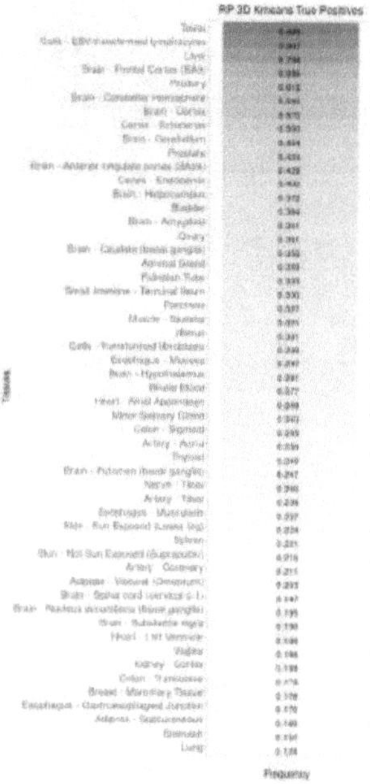
RP 3D Kmeans True Positives
Tissues
Frequency

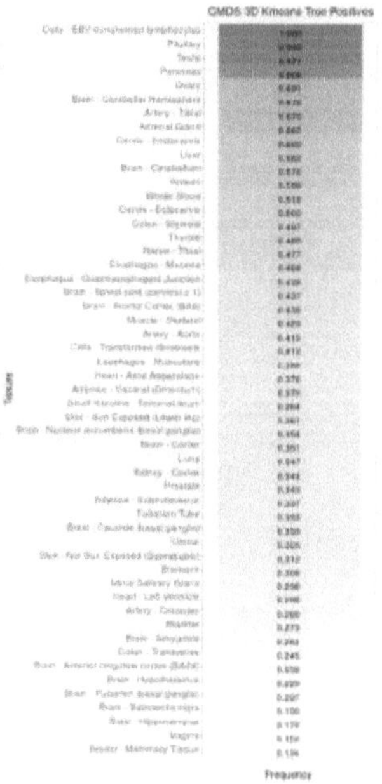
CMDS 3D Kmeans True Positives
Tissues
Frequency

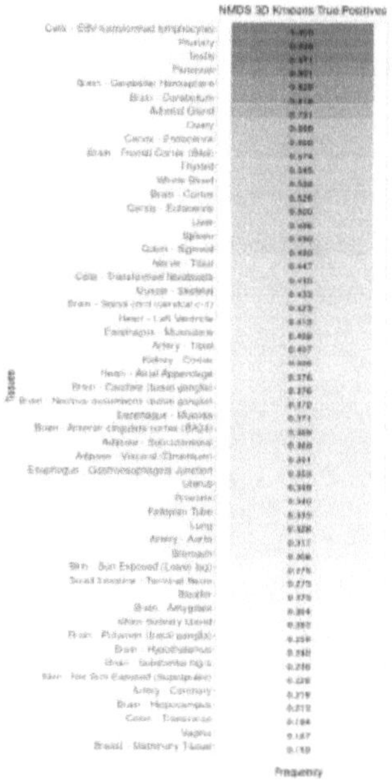
NMDS 3D Kmeans True Positives
Tissues
Frequency

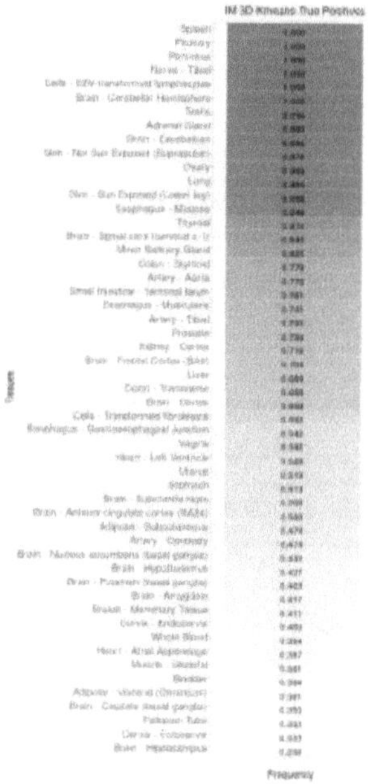
IM 3D Kmeans True Positives
Tissues
Frequency

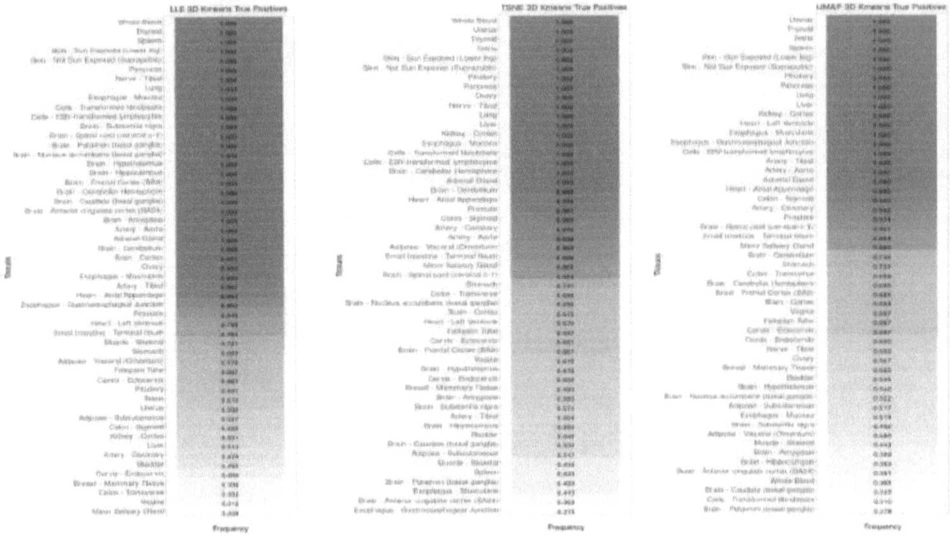

Supplementary Figure 9 - Recall rate based on the clustering evaluation for DR on three dimensions.

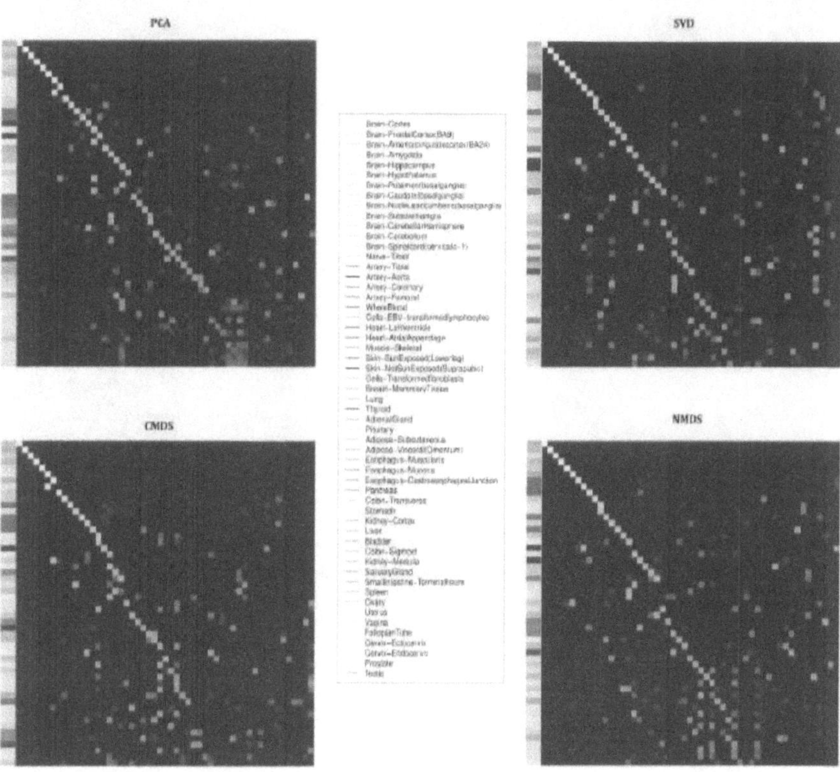

PCA
SVD
CMDS
NMDS
Brain-Cortex
Brain-FrontalCortex(BA9)
Brain-AnteriorCingulateCortex(BA24)
Brain-Amygdala
Brain-Hippocampus
Brain-Hypothalamus
Brain-Putamen(basalganglia)
Brain-Caudate(basalganglia)
Brain-Nucleusaccumbens(basalganglia)
Brain-Substantianigra
Brain-Cerebellarhemisphere
Brain-Cerebellum
Brain-Spinalcord(cervicalc-1)
Nerve-Tibial
Artery-Tibial
Artery-Aorta
Artery-Coronary
Artery-Femoral
WholeBlood
Cells-EBV-transformedlymphocytes
Heart-LeftVentricle
Heart-AtrialAppendage
Muscle-Skeletal
Skin-SunExposed(Lowerleg)
Skin-NotSunExposed(Suprapubic)
Cells-Transformedfibroblasts
Breast-MammaryTissue
Lung
Thyroid
AdrenalGland
Pituitary
Adipose-Subcutaneous
Adipose-Visceral(Omentum)
Esophagus-Muscularis
Esophagus-Mucosa
Esophagus-GastroesophagealJunction
Pancreas
Colon-Transverse
Stomach
Kidney-Cortex
Liver
Bladder
Colon-Sigmoid
Kidney-Medulla
SalivaryGland
SmallIntestine-TerminalIleum
Spleen
Ovary
Uterus
Vagina
FallopianTube
Cervix-Ectocervix
Cervix-Endocervix
Prostate
Testis

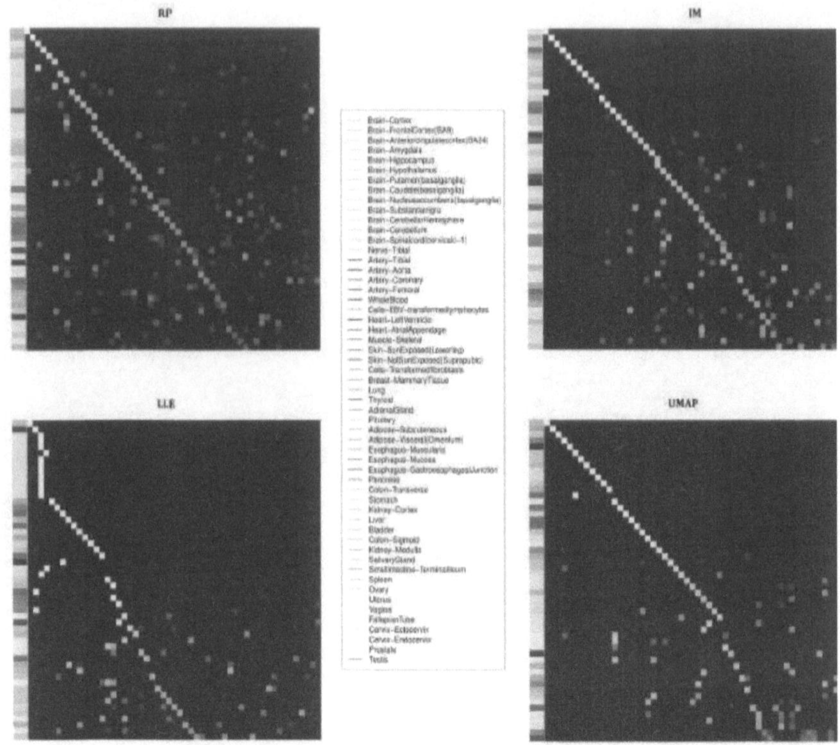

Supplementary Figure 10 - Distribution of percentage of samples per tissue by the assigned clusters in four-dimensional data based on the k-means clustering. Tissues are ranked in descendent order from top to bottom by the frequency of correct assignments to their top cluster.

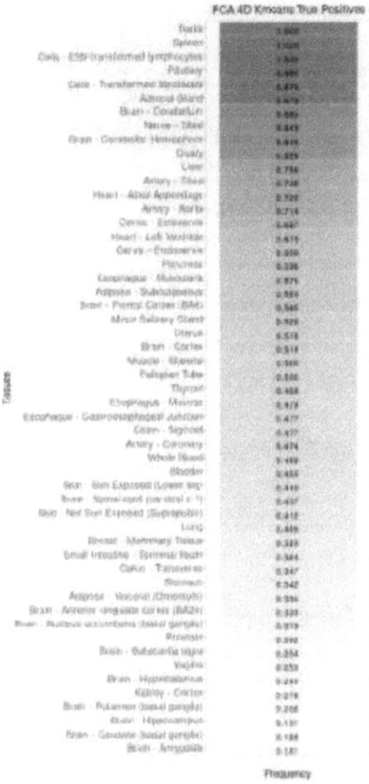
PCA 4D Kmeans True Positives
Tissues
Frequency

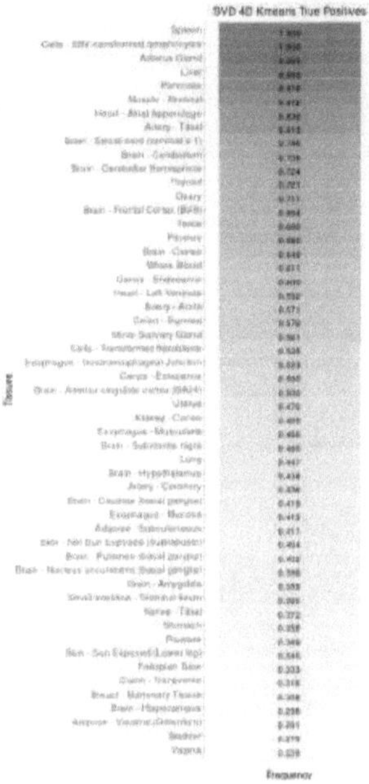
SVD 4D Kmeans True Positives
Tissues
Frequency

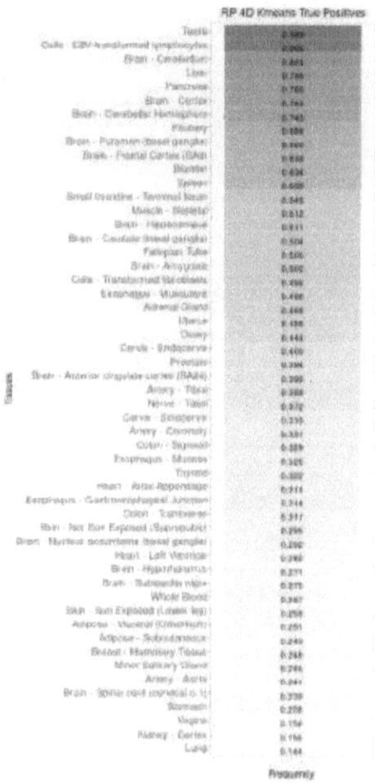
RP 4D Kmeans True Positives
Tissues
Frequency

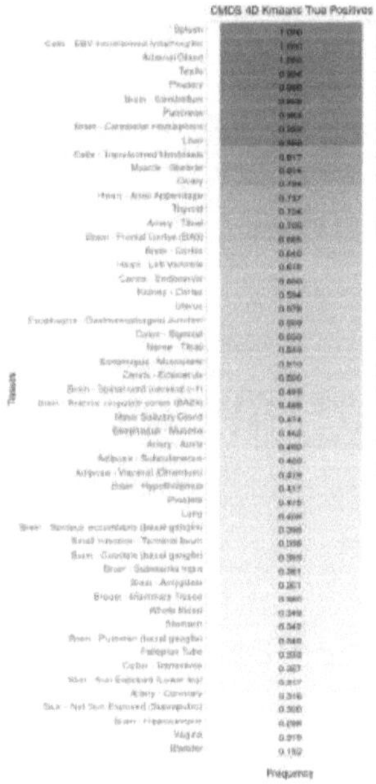
CMDS 4D Kmeans True Positives
Tissues
Frequency

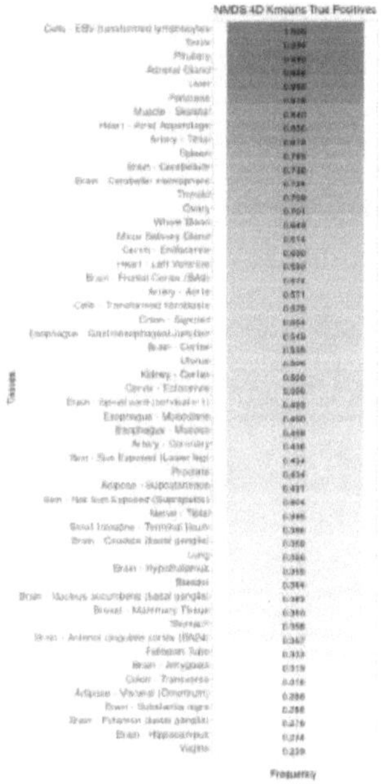
NMDS 4D Kmeans True Positives
Tissues
Frequency

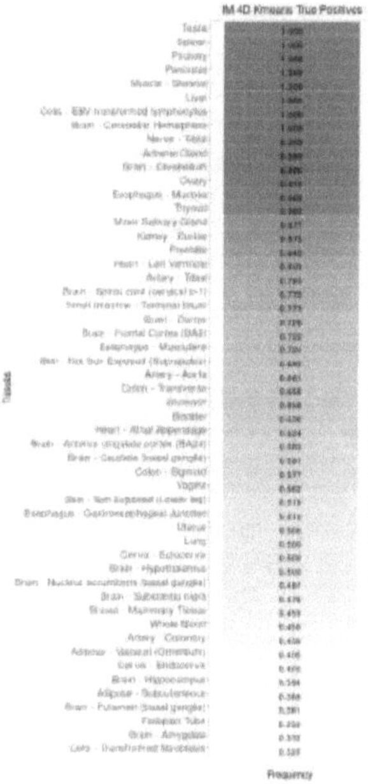
IM 4D Kmeans True Positives
Tissues
Frequency

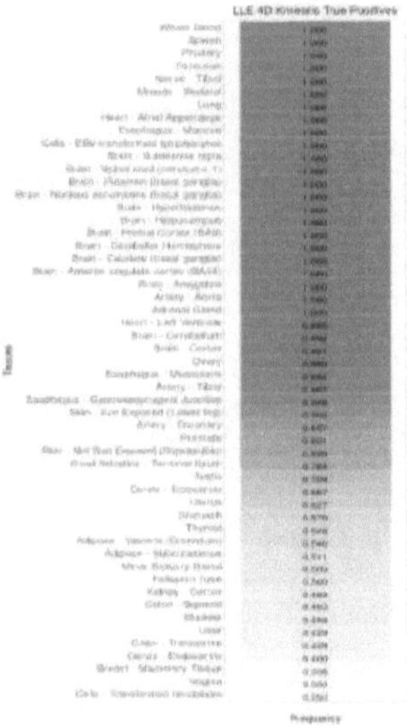

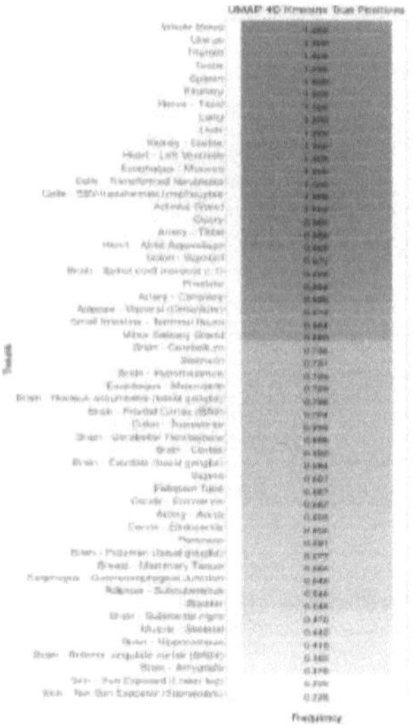

Supplementary Figure 11 – Recall rate based on the clustering evaluation for DR on four dimensions.

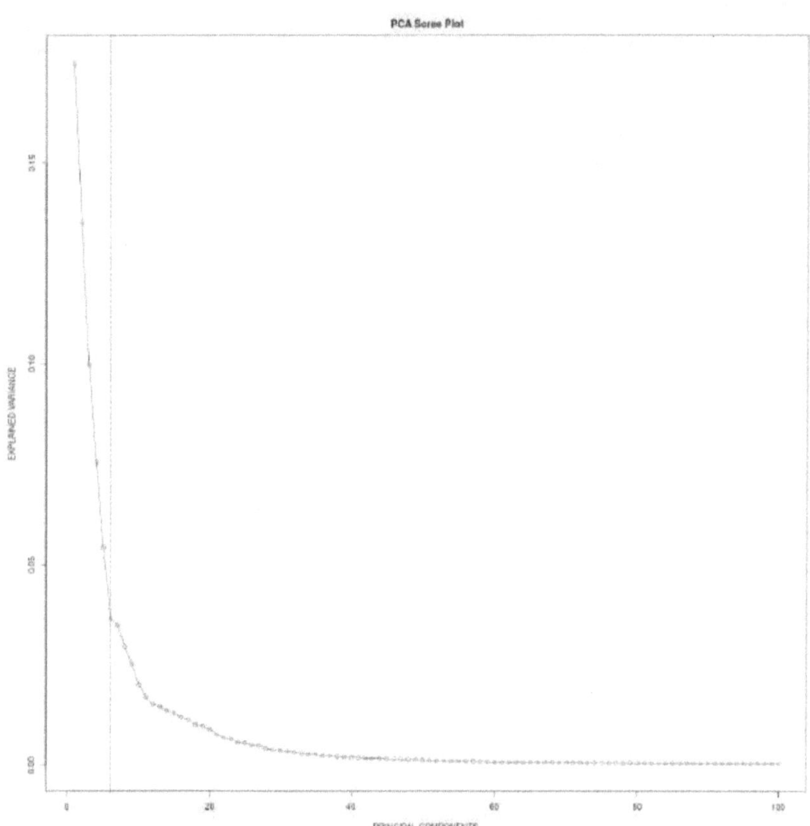

Supplementary Figure 12 - Scree plot of the variance explained for the first 100 principal components of PCA (PCs) of the GTEx dataset. An 'elbow' for this plot can be observed between PCs 6 and 11. Considering the Occam's razor simple of parsimony, 6 PCs were selected for the analysis.

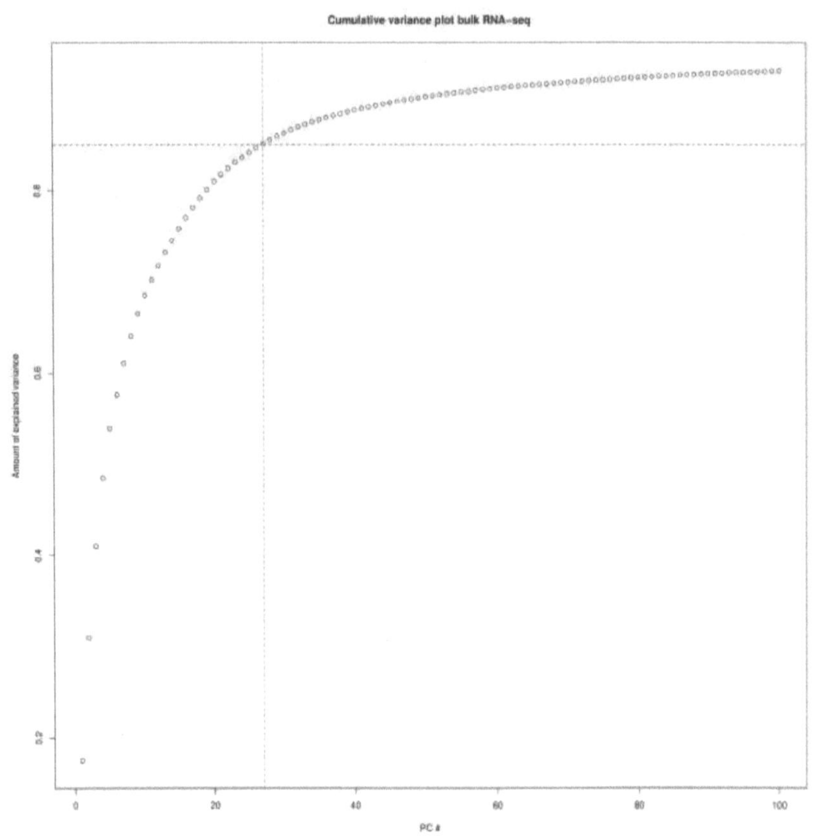

Supplementary Figure 13 - Scree plot of the cumulative variance explained for the first 100 principal components of PCA (PCs) of the GTEx dataset.

Appendix B

Supplementary visualizations of scRNA-Seq dataset

This appendix contains all the supplementary visualizations generated for the analysis of the scRNA-Seq dataset. It comprises the following: a bar plot figure illustrating the number of cells per cell type; Visualization results for all the DR methods of the first and third, first and fourth, second and third, second and fourth, and third and fourth dimensions; Heatmaps of k-means clustering-based evaluation true positives in two, three and four-dimensional reduced data; Heatmaps of the frequency distribution of the cells for each cell type by the assigned clusters for three and four-dimensional reduced data; A scree plot of the variance explained for the first 100 principal components of PCA (PCs); and a scree plot of the cumulative variance explained for the first 100 principal components of PCA (PCs).

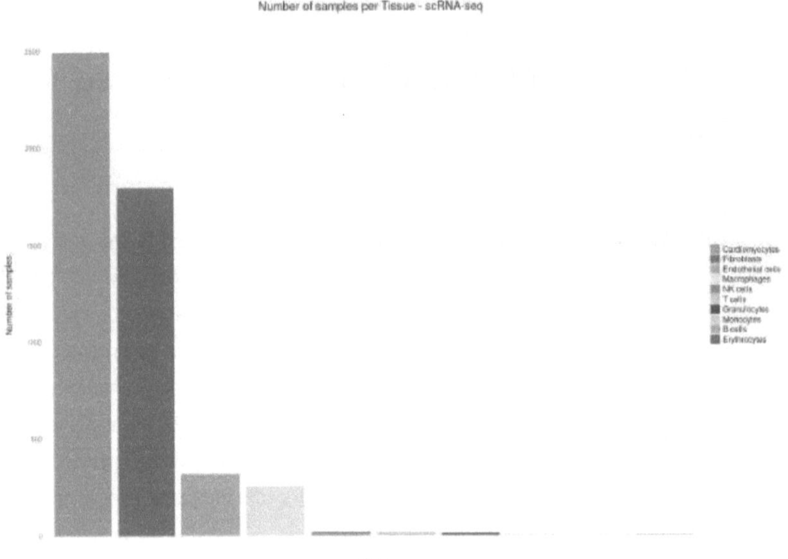

Supplementary Figure 14 - Representation of the number of cells for each cell type of the scRNA-Seq dataset of human embryo cells on descendent order. It varies between 2489 (cardiomyocytes) to 1 (erythrocyte cell). The calculated median number of samples per cell type is 17.

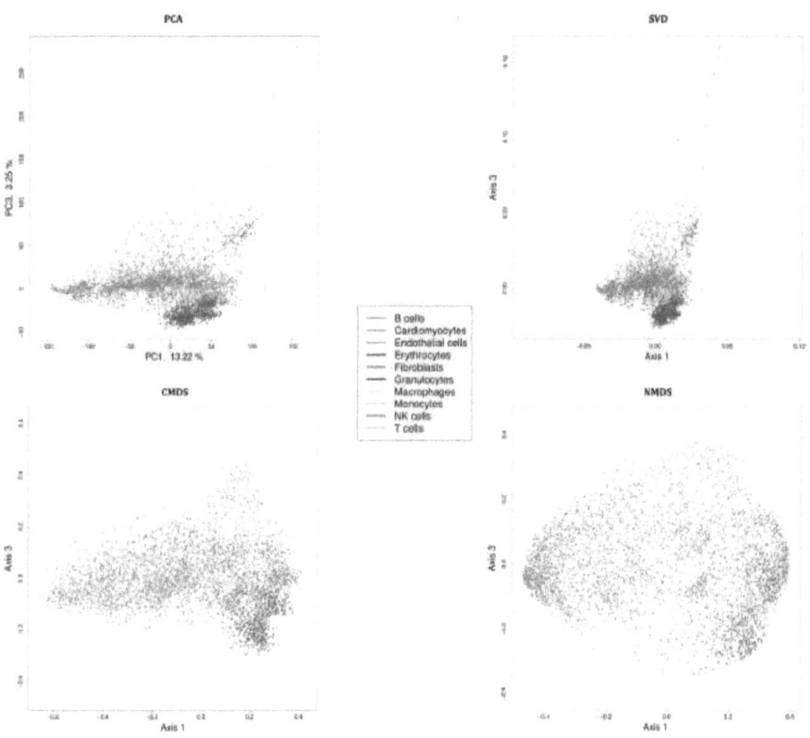

PCA
SVD
CMDS
NMDS
PC3, 3.25 %
PC1, 13.22 %
Axis 3
Axis 1
B cells
Cardiomyocytes
Endothelial cells
Erythrocytes
Fibroblasts
Granulocytes
Macrophages
Monocytes
NK cells
T cells

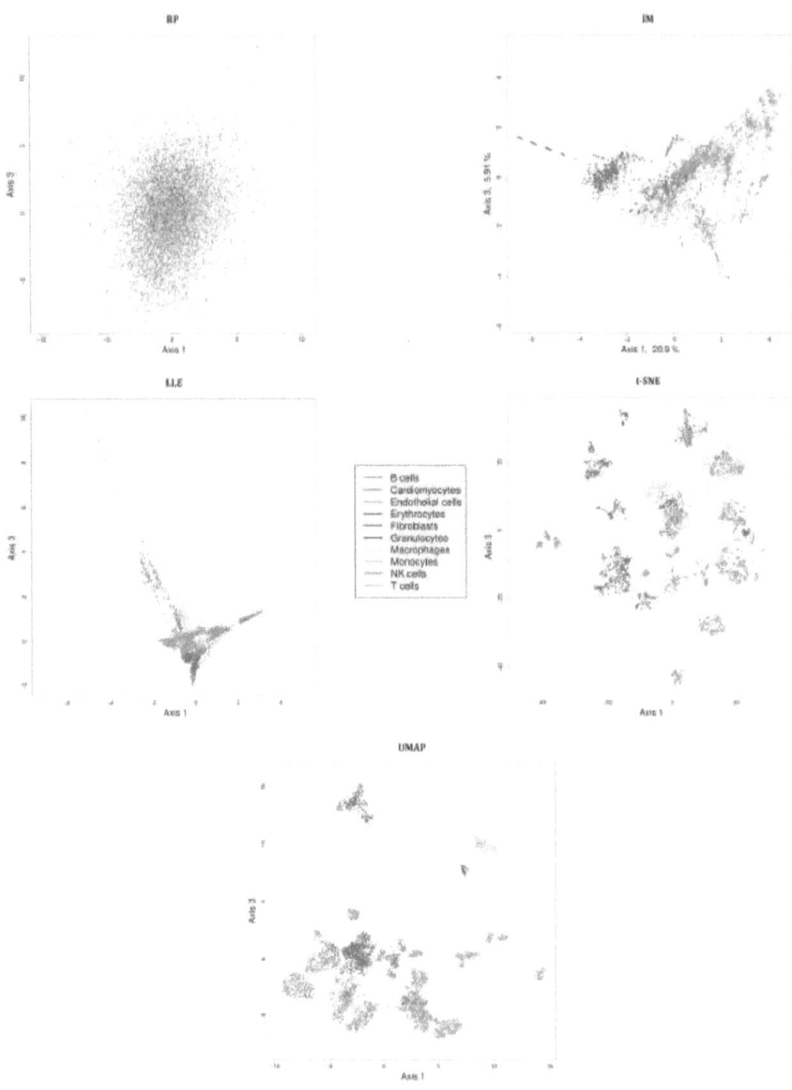

Supplementary Figure 15 - Visualization results for the first and third dimensions on the scRNA-Seq dataset.

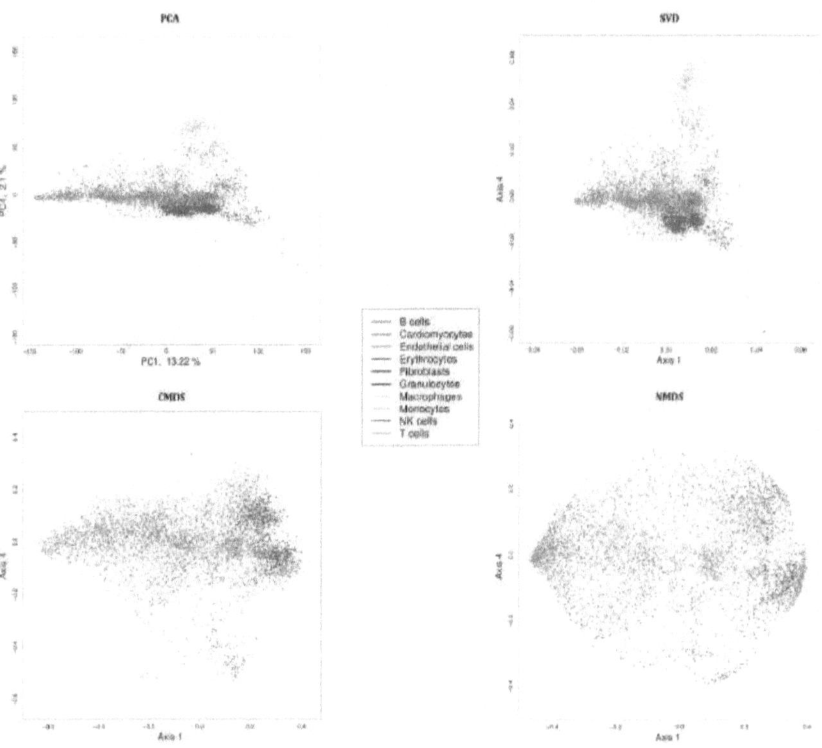

PCA
SVD
cMDS
NMDS
PC1, 13.22 %
PC4, 2.1 %
Axis 1
Axis 4
B cells
Cardiomyocytes
Endothelial cells
Erythrocytes
Fibroblasts
Granulocyte
Macrophages
Monocytes
NK cells
T cells

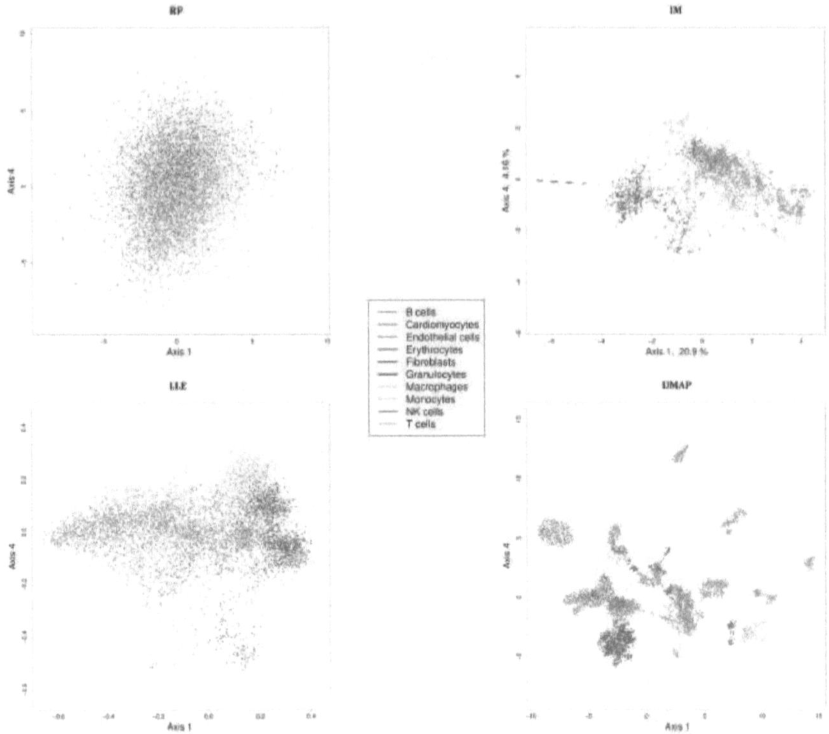

Supplementary Figure 16 - Visualization results for the first and fourth dimensions on the scRNA-Seq dataset.

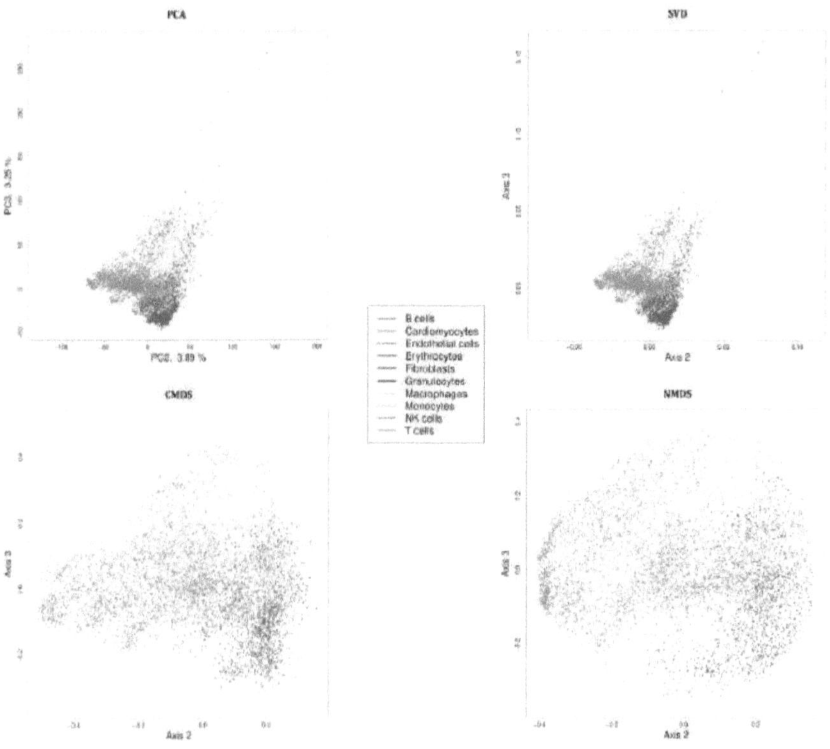

PCA
SVD
CMDS
NMDS
B cells
Cardiomyocytes
Endothelial cells
Erythrocytes
Fibroblasts
Granulocytes
Macrophages
Monocytes
NK cells
T cells

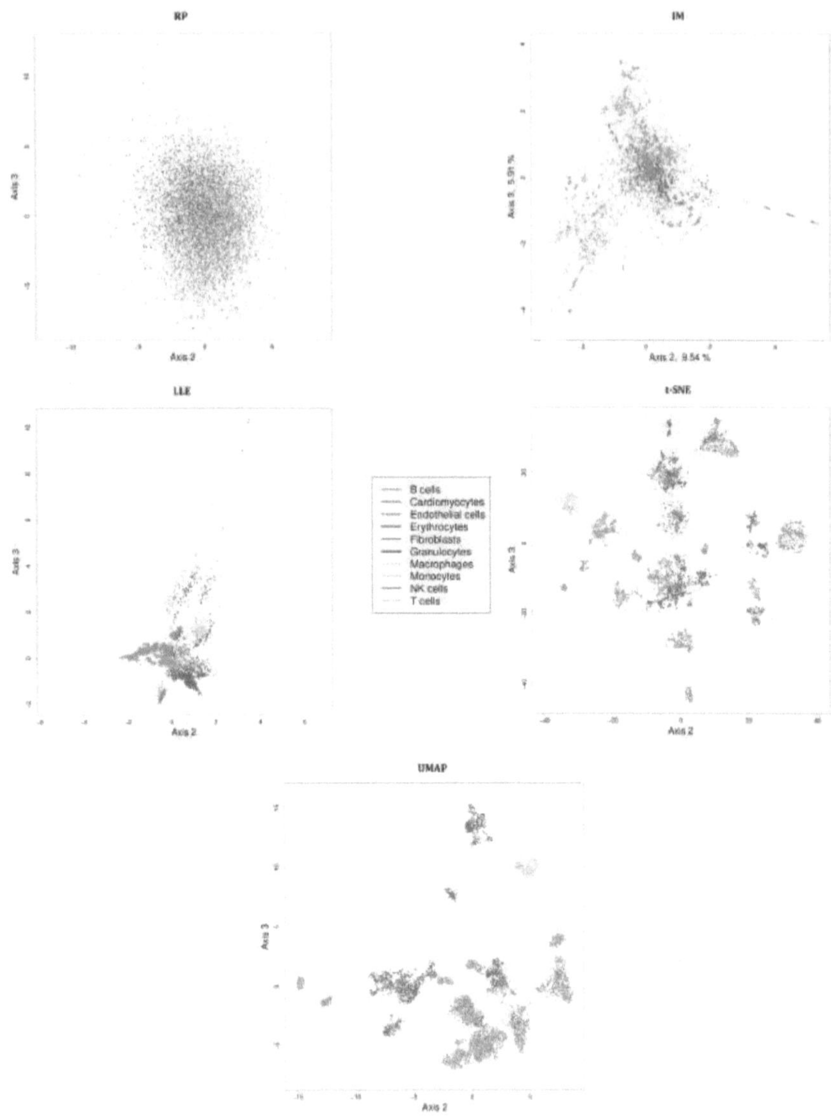

Supplementary Figure 17 - Visualization results of the second and third dimensions on the scRNA-Seq dataset.

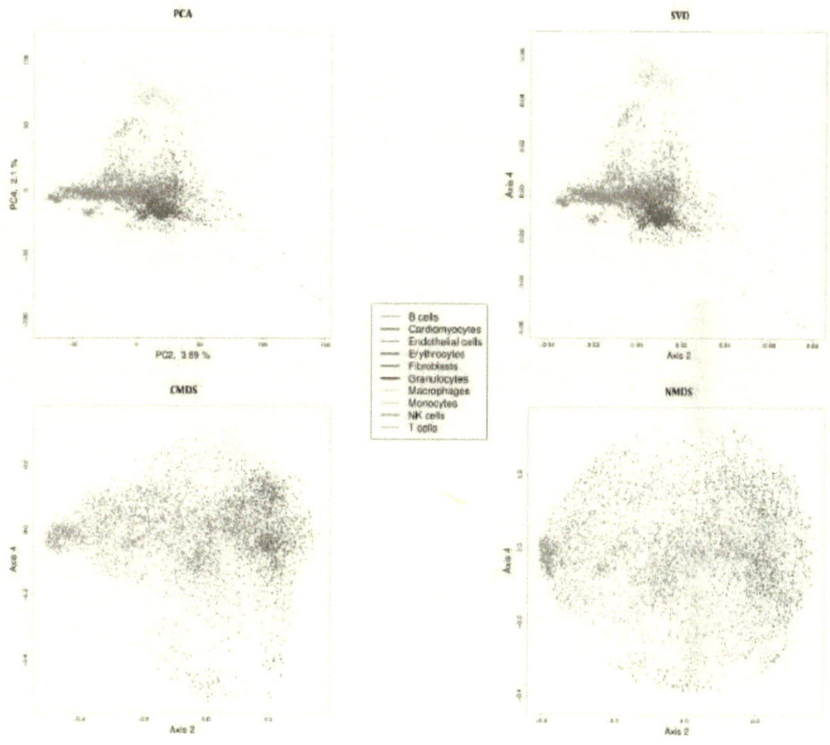

PCA
SVD
CMDS
NMDS
PC4, 2.1 %
PC2, 3.89 %
Axis 4
Axis 2
Axis 4
Axis 2
Axis 4
Axis 2
B cells
Cardiomyocytes
Endothelial cells
Erythrocytes
Fibroblasts
Granulocytes
Macrophages
Monocytes
NK cells
T cells

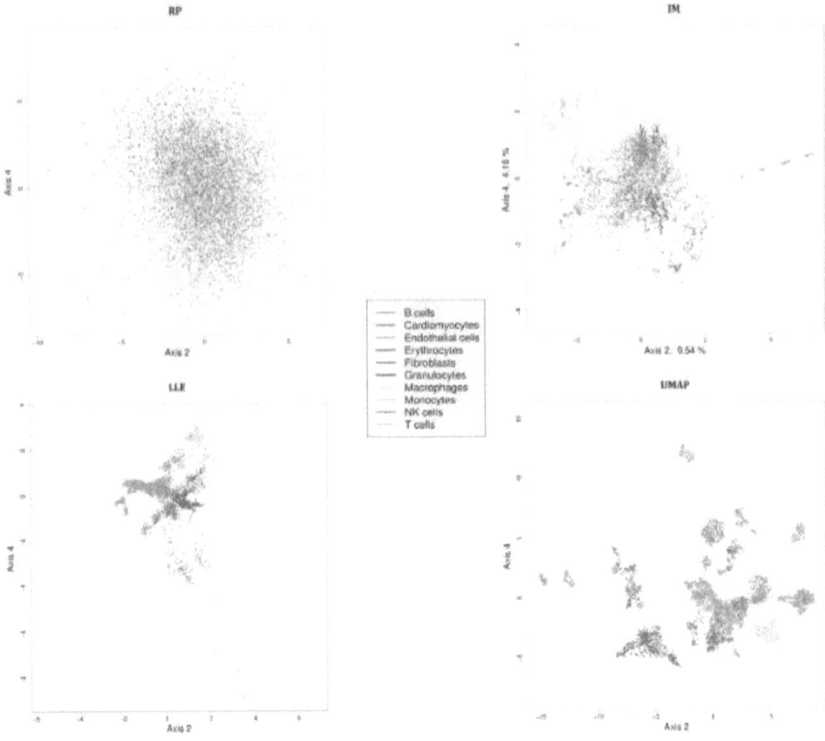

Supplementary Figure 18 - Visualization results of the second and fourth dimensions on the scRNA-Seq dataset.

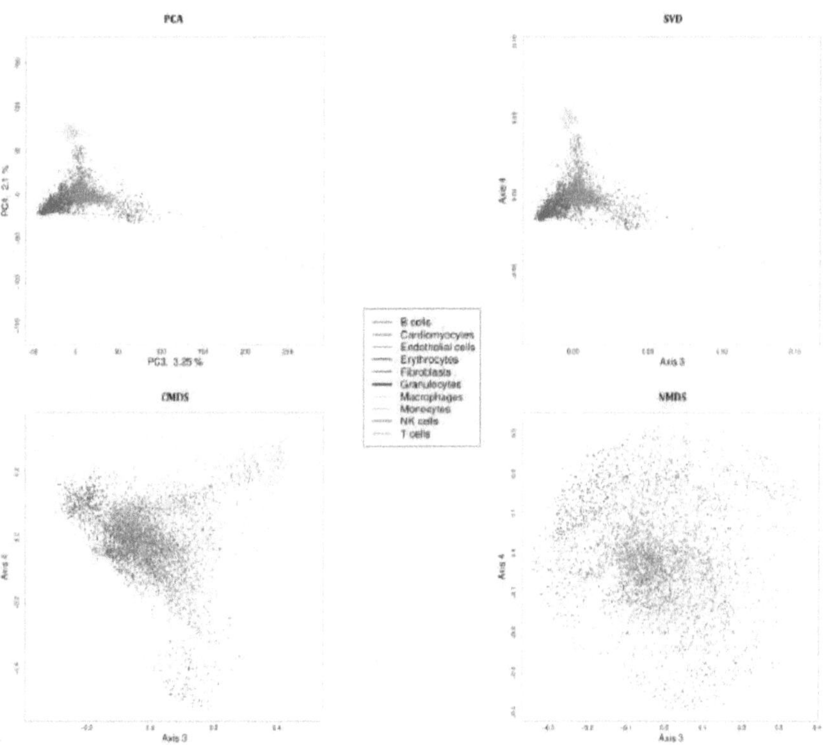

PCA
SVD
CMDS
NMDS
PC4 2.1 %
PC3 3.25 %
Axis 4
Axis 3
Axis 4
Axis 3
Axis 4
Axis 3
B cells
Cardiomyocytes
Endothelial cells
Erythrocytes
Fibroblasts
Granulocytes
Macrophages
Monocytes
NK cells
T cells

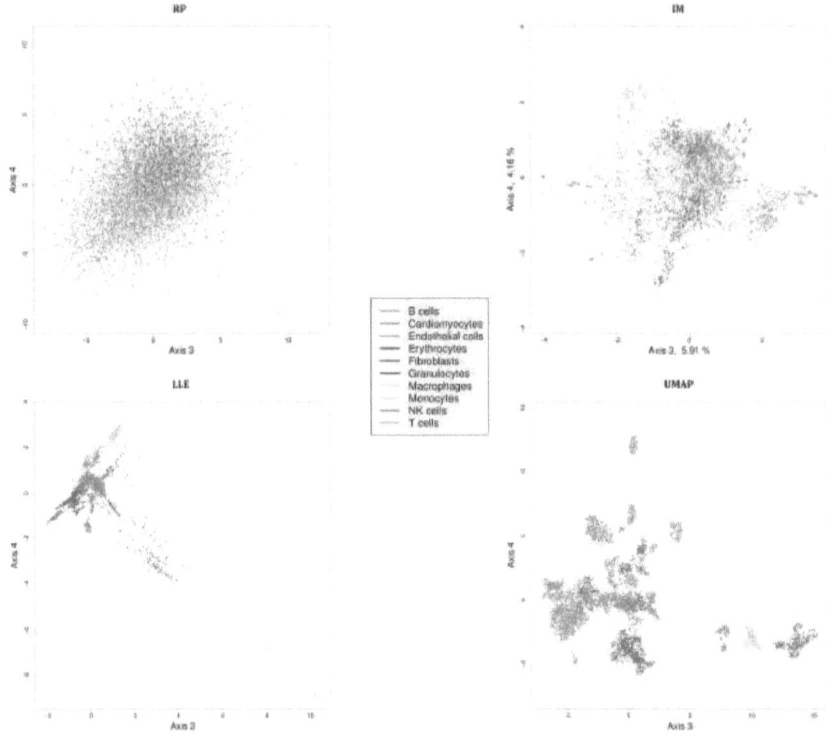

Supplementary Figure 19 - Visualization results for the third and fourth dimensions on the scRNA-Seq dataset.

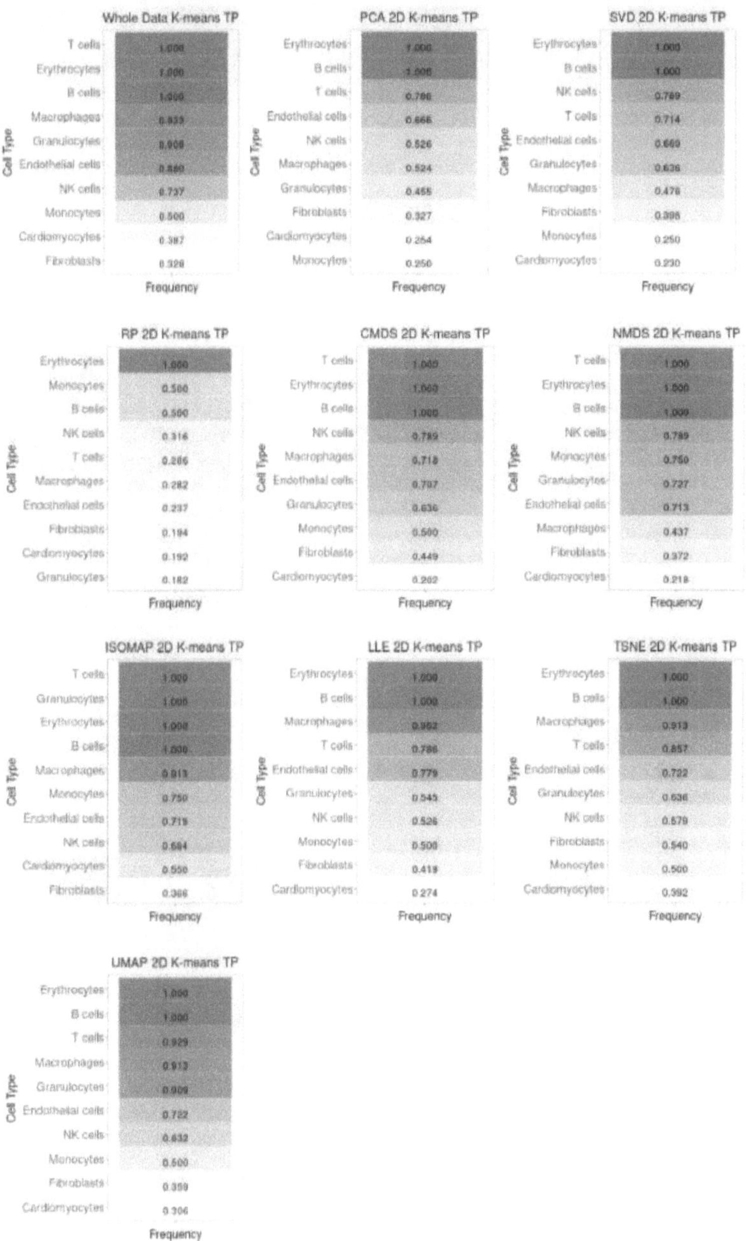

Supplementary Figure 20 - Recall rate based on the clustering evaluation for DR on two dimensions and for the full dataset.

PCA

SVD

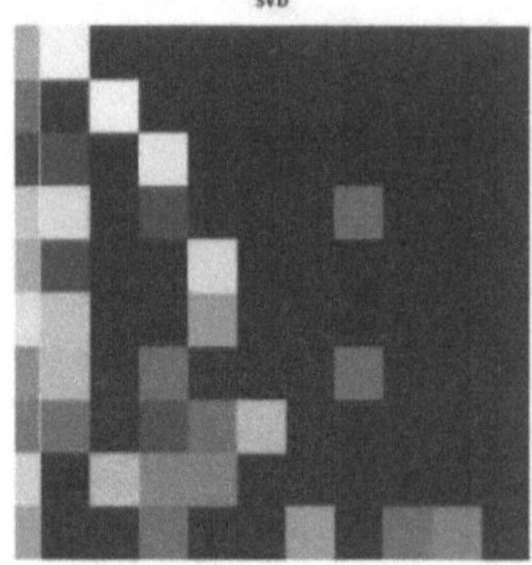

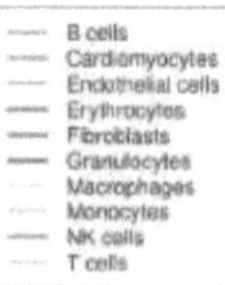

CMDS

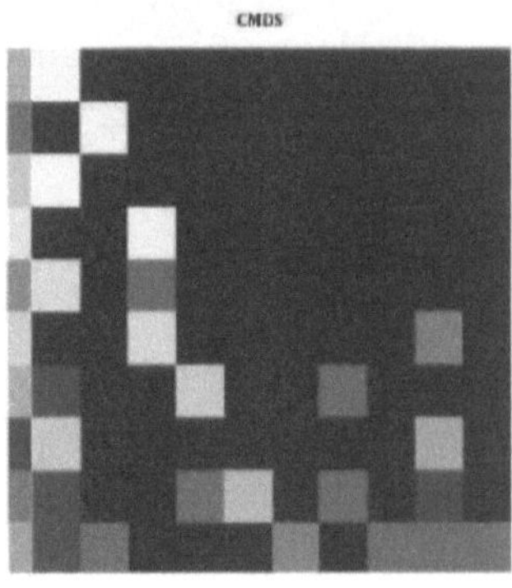

NMDS

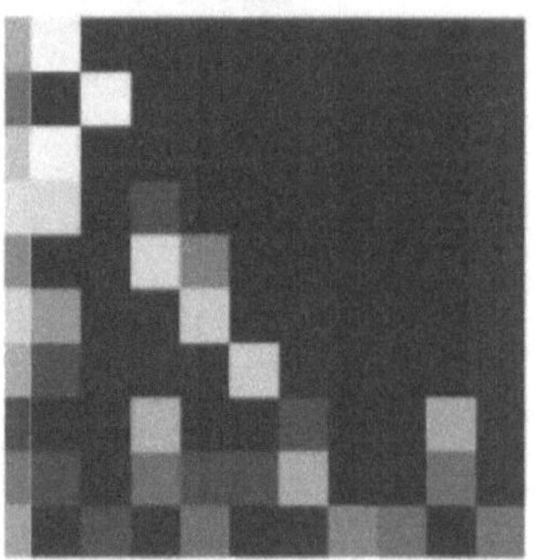

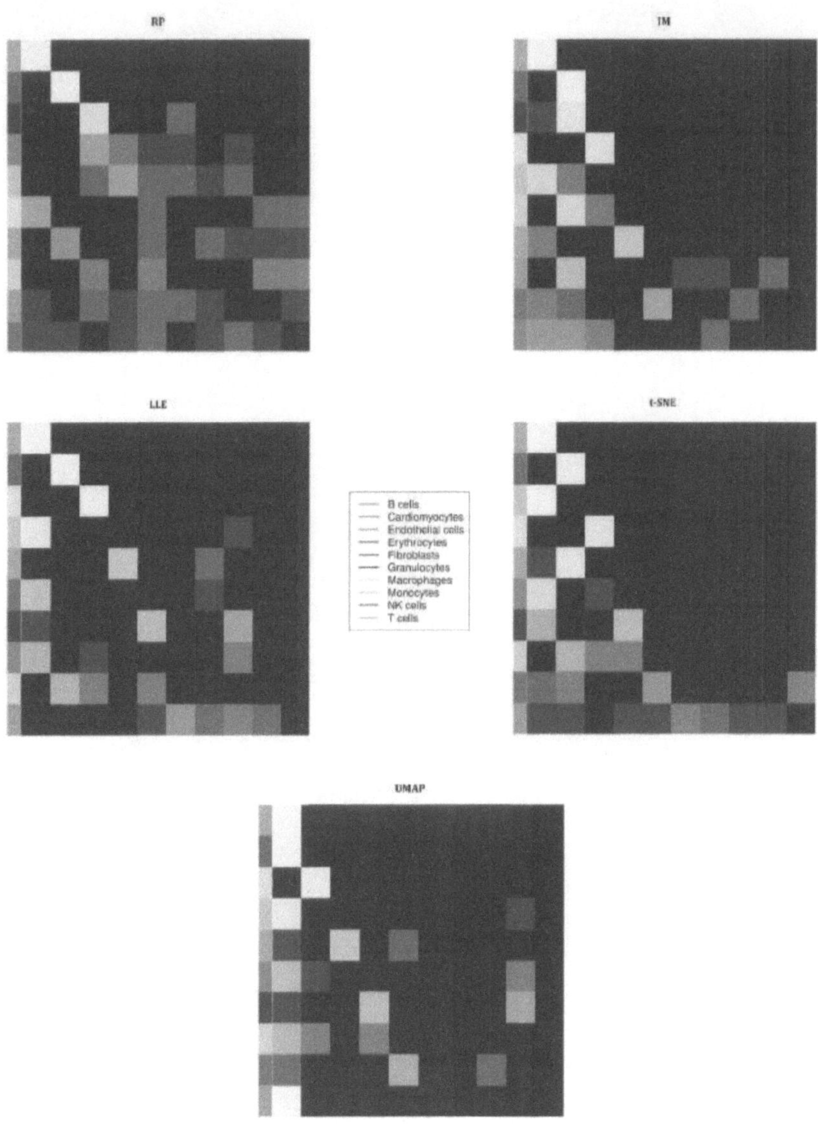

Supplementary Figure 21 - Distribution of percentage of samples per tissue by the assigned clusters in three-dimensional data based on the k-means clustering. Tissues are ranked in descendent order from top to bottom by the frequency of correct assignments to their top cluster.

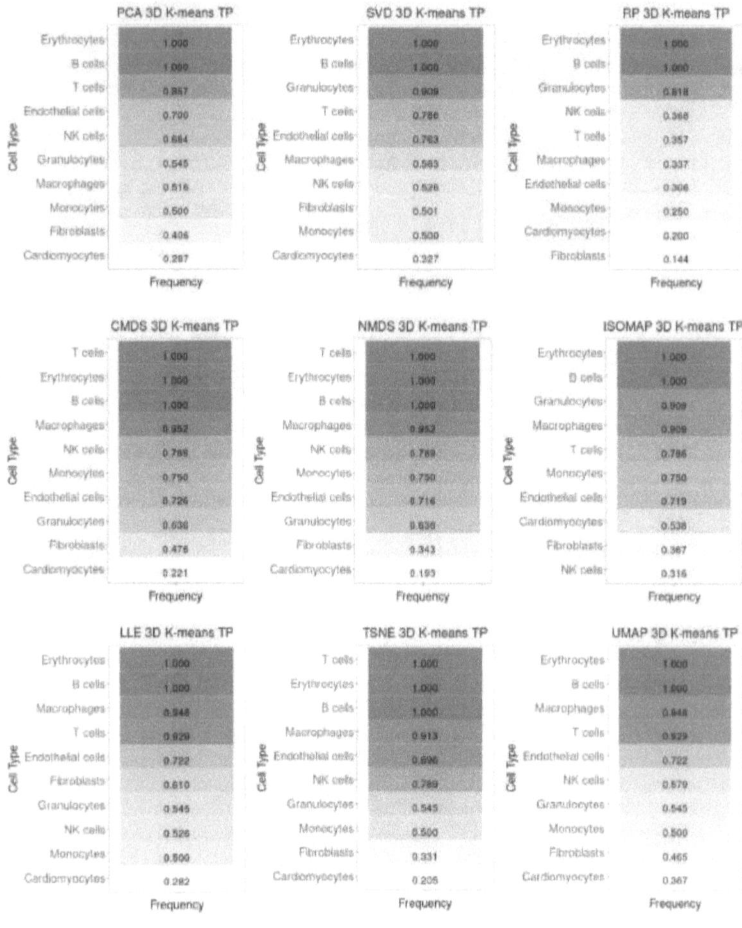

Supplementary Figure 22 - Recall rate based on the clustering evaluation for DR in three dimensions.

PCA

SVD

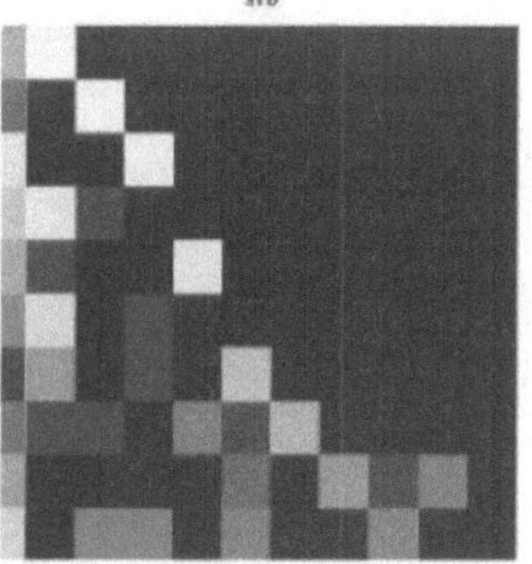

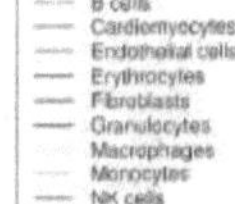

CMDS

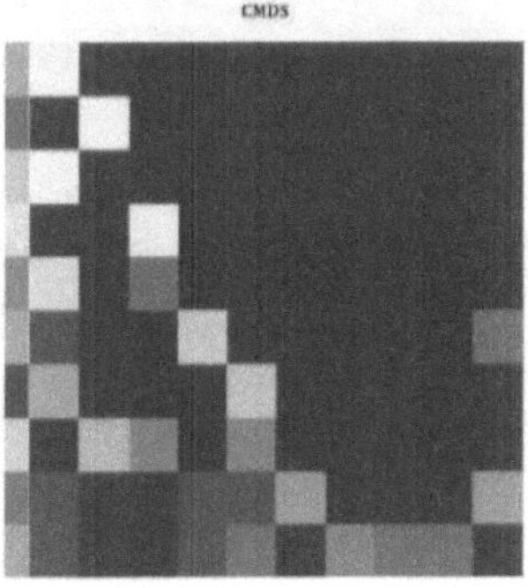

NMDS

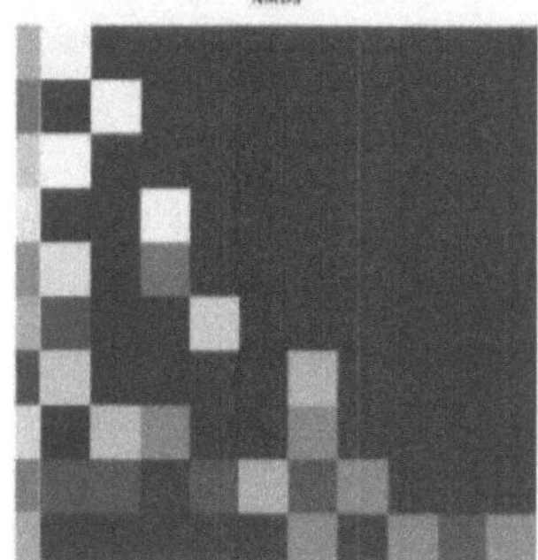

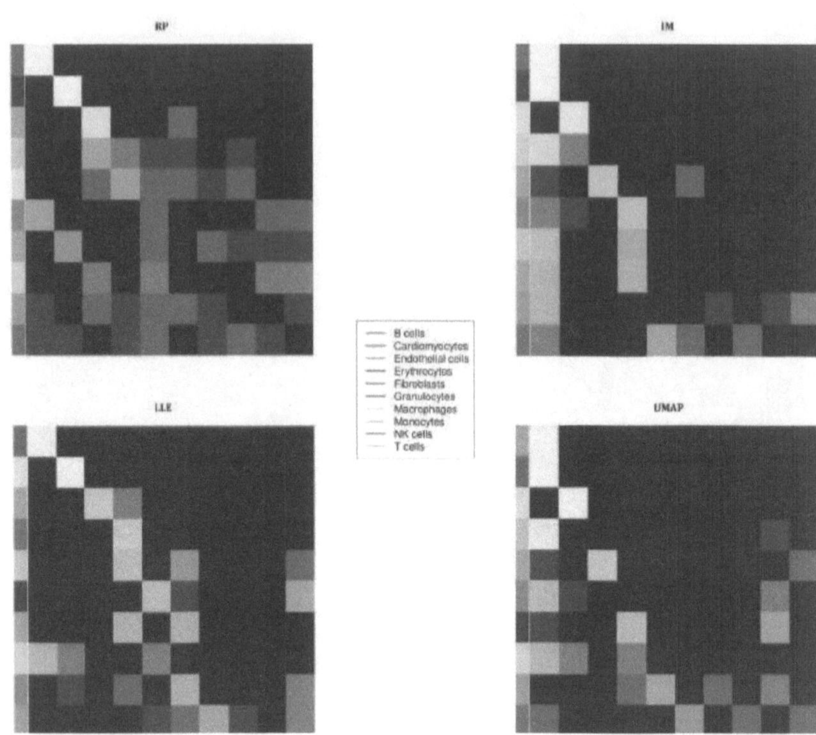

Supplementary Figure 23 - Distribution of percentage of samples per tissue by the assigned clusters in four-dimensional data based on the k-means clustering. Tissues are ranked in descendent order from top to bottom by the frequency of correct assignments to their top cluster.

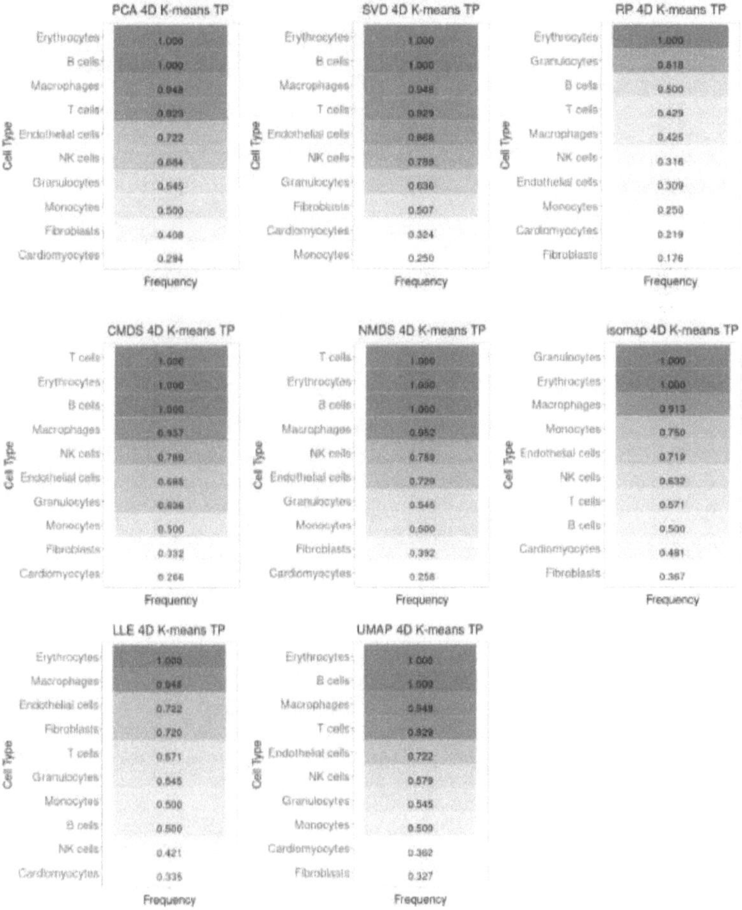

Supplemental Figure 24 - Recall rate based on the clustering evaluation for DR in four dimensions.

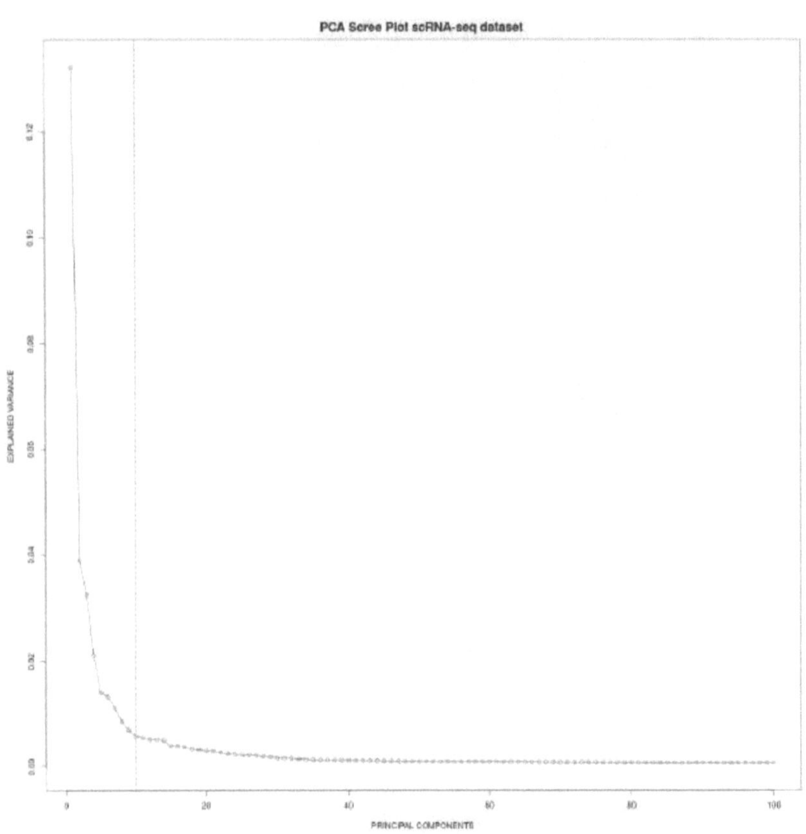

Supplementary Figure 25 - Scree plot of the variance explained for the first 100 principal components of PCA (PCs) of the scRNA-Seq gene expression dataset. An 'elbow' for this plot can be observed between PCs 5 and 11.

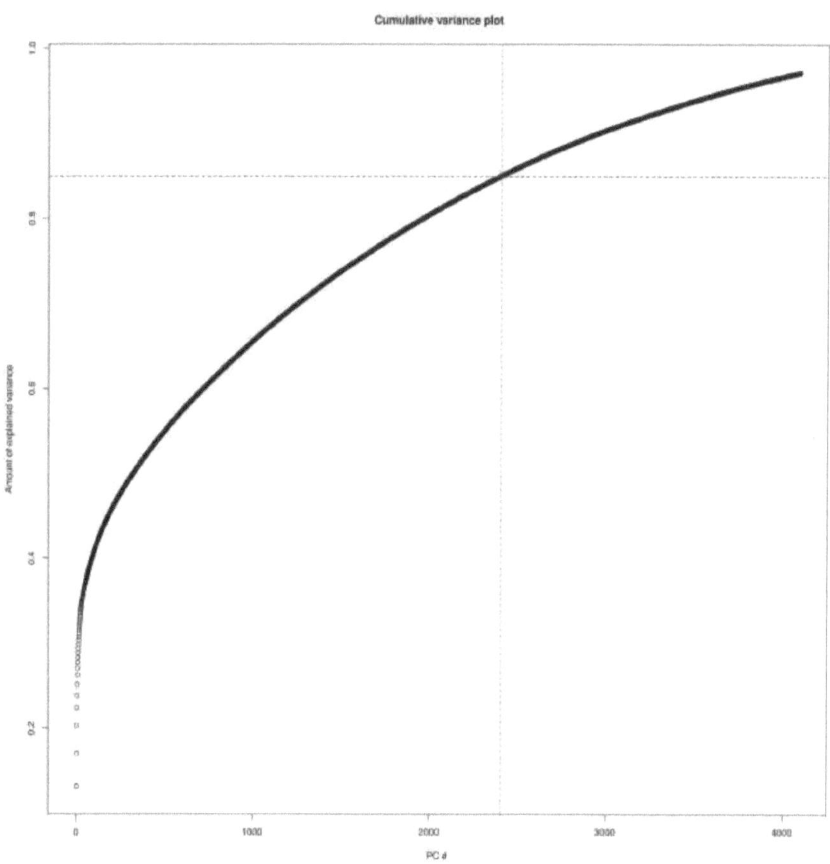

Supplementary Figure 26 - Scree plot of the cumulative variance explained for the first 100 principal components of PCA (PCs) of the scRNA-Seq gene expression dataset. A variance of 85% of this data can be explained by the first 2409 PCs.